D0984164

Scientists
Must Write

Scientists Must Write

A guide to better writing for scientists, engineers and students

Robert Barrass
Principal Lecturer at Sunderland Polytechnic

London

CHAPMAN AND HALL
A Halsted Press Book
John Wiley and Sons, New York

First published 1978
by Chapman and Hall Ltd
11 New Fetter Lane EC4P 4EE

© 1978 Robert Barrass

Filmset in Great Britain by
Northumberland Press Ltd, Gateshead, Tyne and Wear

Printed in the United States of America.

ISBN 0 412 15440 4 (cased edition)
ISBN 0 412 15430 7 (Science Paperback)

Distributed in the U.S.A. by Halsted Press,
a division of John Wiley and Sons, Inc., New York

Library of Congress Cataloging in Publication Data

Barrass, Robert.
    Scientists must write.

    'A Halsted Press book.'
    Bibliography: p.
    Includes index.
    I. Technical writing.   I. Title.
T11.B37   808'.066'5021   77-18561
ISBN 0-470-99388-X

To Ann

Contents

Contents   ix

    The parts of a research report 131
        The front cover 131
        The title page 132
        The Table of contents 134
        The Introduction 134
        The Materials and methods 134
        The Results 136
        The Discussion 137
        The Summary 138
        The Acknowledgements 139
        The list of references 139
    Project reports and theses 141
        Theses 141
        Project reports 143
        Project assessment 144

13 Preparing a report on an investigation 146
    Preparing the manuscript 146
        Things to check in your manuscript 147
    Preparing the typescript 149
        Instructions required by the typist 149
        Things to check in your typescript 151
    Preparing the index 152
    Preparing the typescript for the printer 153
        Correspondence with an editor 154
        Check list for referees (and authors) 155
        Copyright 156
        Things to check in the proofs 157
    Summary 158
        *How to prepare a report on an investigation or a paper
        for publication 158*

14 Talking about science 161
    Preparing a talk 161
        Timing 162
        Using a blackboard 162
        Using slides 163
    Delivery 164

    References 166

    Index 169

Acknowledgements

I write not as a grammarian but as a working scientist, knowing how difficult it is to write well and how important it is that scientists and engineers should try to do so.

I thank the following people, who read the typescript for this book, for their help and encouragement: Professor P. N. Campbell, Director of The Courtauld Institute of Biochemistry, The Middlesex Hospital Medical School, University of London; Mr J. Collerton, Head of Department of Humanities, Newcastle-upon-Tyne Polytechnic; Dr G. Evans, Department of Geology, Imperial College of Science and Technology, University of London; Professor M. Gibbons, Department of Liberal Studies in Science, University of Manchester; and my colleagues Mr E. B. Davison, Principal Lecturer, Department of Electrical and Electronic Engineering, and Dr J. B. Mitchell, Senior Lecturer, Department of Biology, Sunderland Polytechnic. Mr D. W. Snowdon prepared the photographs of the drawings reproduced from other sources (as acknowledged in the legends to the figures concerned). Mr D. B. Douglas drew the cartoons.

Acknowledgements

Preface

Some people say that young scientists and engineers should be taught to write so that they can be employed in management and administration. This is true, but they must also be able to write good English if they are to be effective as scientists and engineers. The requirements in scientific and technical writing are the same: clarity, completeness, accuracy, simplicity, etc. (see Chapter 4). In this book, therefore, the word *scientist* means scientist and technologist, and *scientific writing* means scientific and technical writing.

Writing is part of science but many scientists receive no formal training in the art of writing. There is a certain irony in our teaching scientists and engineers to use instruments and techniques, many of which they will never use in their working lives, and yet not teaching them to write. This is the one thing that they must do every day – as students, and as administrators, executives, scientists and engineers.

This book, by a scientist, is not a textbook of English grammar. Nor is it just one more book on how to write a technical report, or a thesis, or a paper for publication. It is about all the ways in which writing is important to students and working scientists and engineers in helping them to remember, to observe, to think, to plan, to organize, and to communicate.

Chapters 1, 2 and 3 are about all the ways in which writing is important to a scientist or technologist, and Chapter 4 is about the characteristics of scientific writing. I hope that this book will help anyone who has difficulty in putting their thoughts into words

(Chapter 5), and that it will cause them to consider the words they use (Chapter 6) and the way they use them (Chapters 7 and 8). In scientific writing, numbers (Chapter 9) and illustrations (Chapter 10) are important, and the preparation of illustrations is usually the first step in writing the results section of a report, thesis or scientific paper (Chapters 12 and 13). A chapter on reading is included (Chapter 11) and one on talking (Chapter 14).

Where appropriate, the advice given is consistent with American (ANSI), British (BS) and International (ISO) Standards (see p. 123) and with the *Guide for the preparation of scientific papers and abstracts for publication* (UNESCO, 1968).

This book may be read *either* as an alternative to a formal course on scientific and technical writing *or* to complement such a course. To help those who require guidance on a particular aspect of writing a detailed list of *Contents* is included and an *Index*. To help all readers and to reduce the number of cross-references, some essential points are repeated in different contexts.

Exercises are included in appropriate parts of the text (see *Exercises* in the *Index*). These are suitable for self instruction; and suggestions are included to help teachers of science, or of scientific writing, who wish to use these and similar exercises in their courses. Examples of unscientific writing and of poor English are accompanied by notes of faults or suggested improvements. Like Gowers (1973), I do not give the source of such extracts but they were written by people who speak English as their mother tongue; some by professors in universities and all by authors of books or contributors to journals.

30 May 1977   Robert Barrass

I

Scientists must write

When asked why we must write, most scientists and engineers think first of the need to communicate. Communication is so important that it is easy to overlook our other reasons for writing. We write as part of our day-to-day work: to help us to remember, to observe, to think, to plan and to organize, as well as to communicate (Table 1). Above all, writing helps us to think and to express our thoughts – and anyone who writes badly is handicapped both in private study and in dealing with others.

By writing we can communicate with people we know who can judge us by everything that they know about us – by our writing and by our conversation, appearance and behaviour. However, when we write to people whom we have never met, they must judge us in the only way they can – by our writing. The way in which a letter of application is written, for example, may be all that an employer needs to indicate that the applicant is not suitable for the job.

Students are judged by their course work (essays, records of experiments, project reports and theses) and by their performance in written examinations. Only by writing well can we give a good account of ourselves as students, as applicants for employment, and as employees (writing letters, instructions, progress reports, articles and reviews, and scientific papers for publication).

Some scientists and engineers recognize the importance of writing in their work. They take trouble with their writing. Others know that they write badly but they do not worry about this. They are mistaken if they believe that writing is not particularly important in science. Still others, because they are satisfied with their writing, write without thinking about the possibility of improvement.

Many people may be encouraged in their belief that their writing is satisfactory by their success in school and college examinations. However, most students would get higher marks in course work and in examinations if they were better able to put their thoughts into words. Only teachers and examiners know how many marks are lost by students who do not show clearly whether or not they understand their work. In schools, many of the most able pupils fail to show their ability. For example, the following comments are from an examiners' report on a scholarship paper.

All answers included much irrelevant information. Even when a diagram was included, a full written description was also given. Looseness of expression indicated a lack of careful thought. The standard of explanation was disappointing ... many candidates had the knowledge but were unable to express themselves. Very few answers were comprehensive. Marks were lost through omission. Even when they knew the answer many candidates had difficulty in bringing facts together in an effective order.

In the universities the students are bright and clever but are deprived because their teachers have neglected to instruct them in the elements of literary expression (Rivet, 1976). Professor Rivet marked 44 essays by arts and science students. All made spelling mistakes. Malapropisms were frequent. Confused syntax was common: dangling participles abounded, plural subjects were attached to singular verbs, 'thus' was used to introduce a new idea, and tenses were changed at random. In punctuation the commonest error was the interchange of full stops and commas (colons and semi-colons have disappeared), but some authors omitted commas entirely, whilst others sprinkled them on the text as though from a watering can, making it difficult to discover the meaning at first reading.

Many students are clever enough to understand their work and yet unable to communicate their knowledge and ideas effectively.

They need help with their writing more than further instruction in their chosen subject.

The need for improvement is also demonstrated in the writing of working scientists and technologists, who presumably do their best work when they are preparing papers for publication. Yet many authors need the help of an experienced editor:

> The most important part of my editorial work consists of trying to help contributors to say clearly and concisely what they have to say.
>
> ... it is now the duty of every University to see that young people are trained better than ever before in expressing themselves lucidly, concisely, and with precision.
>
> *Cambridge University Reporter*, C. F. A. Pantin (1959)

> As I sit editing an article which may actually have something to say, beneath the ingrained verbiage, and as I try to put nouns back into verbs, passives into actives, and to remove 'isms' and 'isations' from nearly everything, I shudder to think of the amount of congested and unclear writing that the social sciences tolerate...
>
> *Only Disconnect*, Bernard Crick (in McIntyre, 1975)

Despite the efforts of editors, many published papers include verbose and ambiguous sentences which indicate that many educated people either do not think sufficiently about what they wish to write, or they are unable to express their meaning clearly and concisely (see Tables 14 and 15, p. 90–3).

All scientists and technologists should accept that writing is part of their work, but the biggest difficulty facing anyone who wishes to improve the standard of scientific and technical writing is that most educated people are content with their writing:

> We are with difficulty persuaded that we have much to learn about language, or that our understanding of it is defective.... The first condition for improvement in the adult's use of language must be to disturb this ludicrous piece of self-deception.
>
> *Practical Criticism*, I. A. Richards (1929)

Many students of science and technology receive no formal instruction in the art of writing; and when they write badly their English may not be corrected. This is why many scientists do not appreciate how important writing is in science, and why they may remain unaware of their shortcomings.

It is not enough to teach scientists about science. We must also help them to be effective as scientists. And there is a certain irony in teaching students of science and engineering to use techniques and instruments, some of which they may never use in their working lives, and yet not teaching them to write – the one thing that they must do every day as students, and as administrators, executives, managers, scientists and engineers.

The requirements in technical writing are the same as in scientific writing: clarity, completeness, accuracy, simplicity, etc. (see Chapter 4).

There was once a time when Science was academic and useless and Technology was a practical art, but now they are so interfused that ... [most people] cannot tell them apart.

Public Knowledge, John Ziman (1968)

Scientific literature is probably no worse than any other kind but it should be better than it is. It is an amazing phenomenon that the scientist who, as a matter of course, conducts his laboratory research with the greatest refinement and highest precision of which science is capable, is so often willing to dash into print without making sure that his statements are clearly expressed. Surely the scientist, of all people, is under obligation to write not only so that he may be understood but so that he *cannot be misunderstood*.   E. H. McClelland (1943)

Science teachers should help in teaching English by telling young scientists why they need to write and how they should write. Children will not appreciate the importance of writing in all their school work if the teacher of English is the only one who corrects errors in grammar.

Young scientists should know, as early as possible in their

careers, that if they write well they will be better students and better scientists. 'All our education depends on the understanding and effective use of English – as does success in so many aspects of adult life' (HMSO, 1975). Drucker (1952) gave this advice on *how to be an employee*:

If you work on a machine your ability to express yourself will be of little importance. But the further away your job is from manual work, and the larger the organisation that you work for, the more important it is that you know how to convey your thoughts in writing and speaking. This ability to express yourself is perhaps the most important of all the skills that you can possess.

The power of rightly chosen words is great, whether these words are intended to inform, to entertain, or to move (Potter, 1969). There is, however, no short cut to better writing. We learn most by practising this art, by considering the comments of our teachers and colleagues or the advice of editors, and by example – by reading the best prose.

2

Personal records

Writing helps you to remember
Most scientists keep a diary to help them to remember what they have to do and what they have done. They follow detailed instructions, prepared by other people, when they need to be reminded of the procedure to be followed in using an instrument or technique. However, a pupil's first use of writing, as an aid to remembering, is in writing complete sentences dictated by a teacher. Later, students make notes in lectures (and while they are reading, see p. 127) and during investigations.

Making notes during a lecture
The kind of notes taken by students will depend upon the way the lecture is delivered. Sometimes they make detailed notes but remember little of what is said. This happens when the lecturer speaks, almost as in dictation, so that nearly everything can be recorded. In other lectures students make few notes but these are carefully selected headings and sub-headings, words, abbreviations, numbers, phrases and sentences which serve as memoranda. It is in such lectures that students learn most: they are not fully occupied in writing and have time to think.

The lecturer's task is not to provide each student with a neat set of notes – by dictating a summary of a textbook – but to provide a digest of the essentials of the subject supported by examples; to discuss problems, hypotheses and evidence; to explain difficult points, concepts and principles; to refer to sources of further information; and to answer questions. In this way the lecturer acts as a pace-maker for the student who, by listening, can move forward more quickly than he could by reading alone. However, students will find it easier to make notes during a lecture if they have done some preliminary reading and if they have understood the earlier lectures in the same course.

In taking notes the first thing to write, at the top of a new page, is the date and then the title of the lecture. In deciding what to write next, most students take their cue from the lecturer. Some lecturers seem to expect their students to write throughout the lecture and this may be desirable if the information is not readily available in the textbooks. Other lecturers prefer students to listen, to understand, and to use their judgement in distinguishing the main points from the supporting detail.

A good lecturer may start by saying how he is going to treat the subject. If the lecture has been well prepared the notes taken by each student will provide an orderly plan of the lecture, similar to that prepared by the lecturer when he was deciding what to say.

Note-taking helps people to remain attentive. Students must either listen to the lecture and then go to their books, or make notes as they follow the lecturer's argument and explanation. Whichever method is adopted, the students should learn during the lecture and should be ready to ask or to answer questions at the end.

Students should consider what materials to take to a lecture. Some use a bound notebook in which they make notes of many different lectures. Then, when they go home, they waste time by copying out these notes. Other students use a separate notebook for each course of lectures so that they do not have to copy the notes later; but this method is cumbersome. Most students, therefore, prefer to carry one loose-leaf pad. They start each new aspect of their work on a new sheet, leave adequate margins, and space their notes, so that they have room for alterations or additions. With loose-leaf pages there is no need to copy notes since each page can be treated separately and kept in the most appropriate file at home, the order of the pages can be changed if necessary, and new pages can be added at any time and in the most appropriate place.

It is a good idea to standardize (A4 paper is a useful size: 210 × 297 mm) and to use paper with wide lines for writing and blank sheets for drawing. Narrow-lined paper is not suitable for notes and essays because there is no space between the lines for minor additions and corrections.

Keeping a record of practical work
Records made during an investigation should not be made on scraps of paper but in a laboratory or field notebook. Like a diary, this is a permanent record of what was done each day (e.g. the amount of each component in any mixture, the method used in all preparations, the arrangements for the standardization of the conditions for the investigation, the instrument numbers, and the temperature and atmospheric pressure if these are relevant). The notebook is also the place for a drawing of the apparatus, for circuit diagrams, and for the readings, descriptions and drawings made during observations.

The organization of data should start as they are recorded, on carefully prepared data sheets. These tables should be pages in the notebook, or they should be securely fixed to the appropriate page. Units of measurement must be given in each column head. Data sheets are an aid to observation. They direct attention to the readings required, so that things are recorded in order and at the right time. They also facilitate the perusal and analysis of data.

Every note should be dated. The date cannot be remembered and it may assume great importance later, indicating the order in which things were done. The date is also the key to records made by other people of such things as climate and day-length and the state of the tides. For the same reasons, keep a record of the starting time, of the time when each note is made during an observation, and of the time the observation ends.

Details such as these will be required if the work is to be repeated and for the *Materials and methods* and the *Results* sections of a report: they should not be trusted to the memory. It may not be possible to write your report if some detail has not been recorded: then all the time and money spent on the work has been wasted. It will also be easier to write the *Introduction* and *Discussion* sections if your reasons for starting the work and notes on the development of arguments and hypotheses are recorded during the investigation.

Notes should not be made in rough and copied later. This wastes time and mistakes may be made in copying. It is best to make neat records and to write in carefully constructed sentences.

If a notebook is lost, after an investigation that has taken weeks, months or years, much time and money has been wasted. Working scientists, therefore, are advised to keep a carbon copy of each page of their notebook, on loose-leaf, and to keep these copies in another place. Those who fail to take this precaution may lose irreplaceable notes (Fig. 1). Such losses, which are most likely to be caused by fire, are never expected. In projects and in research do not dismantle the apparatus until you have completed your observations, analysed your data, and written the first draft of your report. Things which seem unimportant during an investigation may prove to be important later. Your notebook should then provide the information about why, how and when you did things. In your field notebook there should be a similar record not only of why, how and when but also of where you made the observations.

Record any unexpected observations carefully since it is often from these and from experiments that go wrong that we learn most.

I had, also, during many years followed a golden rule, namely, that whenever a published record, new observation or thought came across me, which was opposed to my general results, to make a memorandum of it without fail and at once; for I had found by experience that such facts and thoughts were far more apt to escape from the memory than favourable ones.

Life and Letters, Charles Darwin (1809–82)

Writing helps you to observe
Observation is the basis of science and preparing a description, like making an accurate drawing (Fig. 11A), helps to focus attention

Where did I put my notes?

Fig. 1. *Keep a carbon copy of each page of your notebook in a safe place*

on an object or event. Writing is necessary for precise description and is an aid to learning.

When one point has been covered adequately in a description, look for something else to describe. This will help you to see more and your description will not then be confined to the most obvious things. As a trained observer you should try to miss nothing.

The revision of a description provides an opportunity for the rearrangement of observations: so that there is a clear distinction between what seem to be the most conspicuous features and what is detail relating to these, so that events are described in the correct time sequence or in some other logical order, and so that attention is drawn to different observations which seem to be related.

Writing helps you to think

We think in words, and in writing we try to capture our thoughts. Writing is therefore a creative process which helps us to sort our ideas and preserve them for later consideration.

Hardly any original thoughts on mental or social subjects ever make their way among mankind, or assume their proper importance in the minds even of their inventors, until aptly selected words or phrases have, as it were, nailed them down and held them fast.　　*A System of Logic*, John Stuart Mill (1875)

... the toil of writing and reconsideration may help to clear and fix many things that remain a little uncertain in my thoughts because they have never been fully stated, and I want to discover any lurking inconsistencies and unsuspected gaps. And I have a story.　　*The Passionate Friends*, H. G. Wells (1913)

..., an English course consisting only of grammar would be very barren, and command of language is best obtained by using it as a vehicle for disciplining and recording thought and stimulating imaginative thinking.

The Language of Mathematics, F. W. Land (1975)

Preparing an essay or report makes you set down what you know and helps you to recognize gaps in your knowledge, and so leads you to a deeper understanding of your work.

English is not like other school subjects: it is the condition of academic life. More than this, for English speaking people, English is the only means of expression by which they become articulate and intelligible human beings (Sampson, 1925). When, therefore, someone says 'I'm no good at English' what he or she really means is ... 'I'm no good at thinking straight, I can't talk sense, I'm no good at being myself' (Strong, 1951). The teaching of English is therefore the point at which all education must start, with every teacher a teacher of English.

Even educated people write badly when they have not thought sufficiently about what they wish to write.

Words are ... the only currency in which we can exchange thought even with ourselves. Does it not follow, then, that the more accurately we use words the closer definition we shall give to our thoughts? Does it not follow that by drilling ourselves to write perspicuously we train our minds to clarify their thoughts.

On The Art of Writing, Sir Arthur Quiller-Couch (1916)

Writing helps people to arrange their thoughts and to plan their work (see Chapter 4: *Thinking and planning*, p.37).

The effort of formulating hypotheses (putting into words the possibilities we envisage) results in a spelling out to which we may later return in the light of further experience and in search of further possibilities. The hypotheses draw our attention to further questions and to fresh hypotheses. Our statements form a basis for contemplation and a spur to further investigations (HMSO, 1975).

The value of writing in study is indicated by the use lecturers make of the essay, by the contribution of the written report in project assessment, and by the place of the thesis in a student's preparation for a higher degree examination. One of the duties of a university is to give instruction 'but it is our higher function to teach our students to think, and of this accomplishment the thesis or essay is the chief evidence' (Allbutt, 1923).

Writing an account of an experiment

The account must be based on records prepared during the investigation (including notes of dates and times, of materials and methods, and data). The headings used, after an appropriate title, are: *Introduction*, *Materials and methods*, *Results*, *Discussion and conclusions* and *List of references*. If other headings are more appropriate they should be used, except that students may be asked to use the same format for all reports.

The title should indicate what the experiment was about. The *Introduction* should state why the experiment was performed. The *Materials and methods* section should be sufficiently detailed so that someone with a similar background and training could repeat the observations and obtain similar data. The *Results* section is a statement of what was observed and it includes the results of the analysis of the data (as tables, graphs or statistics included in the text). The data (or representative data) may also be required. The *Discussion* is your interpretation of your results but relevant published work by other people may also be mentioned. The *Conclusions* should be listed, with each conclusion a separate numbered statement.

Thinking and remembering

It is a good idea to carry a pencil and a pocket notebook so that otherwise fleeting thoughts can be recorded. Make a note of ideas and associations, and of possible further investigations, as they come to mind, so that these may be considered later and not forgotten.

Progress reports

In project work and in research, writing should be part of the investigation and not an unwelcome task to be undertaken at the end. Science and writing are not separate and successive tasks. It is best to write before the investigation to help to define the problem and to plan the methods and experiments (see *Preparing the manuscript*, p. 146). Those who do not start to write until the end of an investigation make their work unnecessarily difficult. An outline prepared at the start should be revised and augmented as the work proceeds. This provides practice in the art of writing and helps to draw attention to any further work that may be needed. In this way, as information is added under each heading, the draft report remains an up-to-date record of progress.

A useful aid in getting a clear understanding of a problem is to write a report on all the information available. This is helpful when one is starting on an investigation, when up against a difficulty, or when the investigation is nearing completion. Also at the beginning of an investigation it is useful to set out clearly the questions for which an answer is being sought. Stating the

problem precisely sometimes takes one a long way toward the
solution. The systematic arrangement of the data often discloses
flaws in the reasoning, or alternative lines of thought which had
been missed. Assumptions and conclusions at first accepted
as 'obvious' may even prove indefensible when set down clearly
and examined critically.
The Art of Scientific Investigation, W. I. B. Beveridge (1968)

A progress report is of value to the supervisor and to those supporting the research, but it is of greatest value to the writer. It helps
the scientist to plan additional observations, to avoid irrelevant
and therefore time-wasting distractions, to see the project as a
whole, and to recognize when the work is complete.

Table 1. What scientists write

Private records

Laboratory or field notes; diaries; case histories
Data sheets
Descriptions as an aid to observation
Notes of lectures and from publications
Index cards
Notes of ideas and memoranda; pocket notebook
Notes of information and ideas as an aid to thinking and planning
Notes for lectures

Communications

Postcards, letters and memoranda
Essays, articles, pamphlets and books
Instructions
Technical reports, descriptions and specifications
Progress reports
Theses (Dissertations)
Research papers for publication
Press releases
Book reviews

Practise writing

1 Keep a record of all your practical work in a laboratory notebook.

2 Use writing and drawing as aids to observation and description. Much scientific writing is based on clear and accurate description; but if we ask for descriptions of an event, a process, or a thing, from a class of students for example, we shall find considerable differences, not only in the quality of the writing but (most alarmingly) in the perceptions of the thing described (Henn, 1961).

3 In project work, write an account of your observations and experiments as you go along. Think of the report as part of the investigation.

4 Use writing as an aid to thinking.

Teachers may start a course on scientific writing by asking their students to prepare, as a basis for discussion, a list of the kinds of writing undertaken by scientists (Table 1). If possible, teachers (including teachers of English) should base their instruction on the students' practical notebooks, reports and essays; and instructors and editors in industrial and other establishments should use letters, memoranda and reports, produced in their own organization. This is the best way, and perhaps the only way, to demonstrate to scientists and engineers that if they write well they will be more effective in their work.

3

Communications

When you write reports for administrators and politicians, or write letters, or try to popularize science (all tasks that require tact, imagination and an understanding of the reader's needs) you must write clearly and forcefully, in words that educated people will understand, so that your conclusions or recommendations stand out from any supporting detail, and so that your message cannot be misunderstood.

Internal reports
You are judged by your writing, and your value as an employee depends not only upon your knowledge of science or engineering but also on your ability to communicate information and ideas.

> The popular picture of the engineer, for instance, is that of a man who works with a slide rule, T-square, and compass. And engineering students reflect this picture in their attitude toward the written word as something quite irrelevant to their jobs. But the effectiveness of the engineer – and with it his usefulness – depends as much on his ability to make other people understand his work as it does on the quality of the work...
>
> *How to be an Employee*, Peter F. Drucker (1952)

You must communicate with the people you work for and, as you take on the responsibilities of leadership, you must pass on clear instructions to others.

It is not enough for you to have a good idea or to do good work; you must also be able to make other people understand what you are doing, why you are doing it, and with what result (see *Progress reports*, p. 13). It is easy to make a complicated subject seem complicated but intelligence and effort are needed if the information and ideas are to be presented as simply as possible. The most important things to consider, as in all communications, are: why is the report required, what information is required, and by whom (see Chapter 8).

An internal report should conform to the house rules or to the requirements of your head of department. If he gives no guidance, try to arrange all the information he requires in an effective order, on one page, with an appropriate heading and sub-headings, and with your conclusions or recommendations at the end. For such a short report, any supporting details, graphs or diagrams should be listed at the foot of the page and included on separate sheets attached to your report. (For suggestions on the presentation of longer reports see Chapter 12).

Many young people, after qualifying as scientists, go directly into administration or start training for management: their success depends upon their ability to communicate with people inside and outside their own organization. The administrator who takes the trouble to improve his writing will be more effective as an administrator: his work will be better organized, easier to read, and easier to understand; and he is more likely to consider the needs and feelings of others.

Writing letters and memoranda

A letter is a good test of your ability to communicate effectively. Every letter is an exercise in public relations; it represents you and usually also your employer. You should therefore take care over the appearance, layout and content of every letter to ensure that it makes a favourable impression on the recipient. The organization of all except the shortest letters can be improved, and their length reduced, if you make notes of the points you wish to emphasize and then number these in an effective order, before you write or dictate a letter. Important letters should be read by someone else in the same organization and/or checked by the writer on the following day, and then – if necessary – they should be revised. Apart from such necessary delay, all correspondence should be dealt with promptly as a matter of courtesy and to increase efficiency. In writing a letter or memorandum the basic requirements are the same as in other communications. You must know what you wish to say: then convey this message clearly, concisely and courteously, appreciating the reader's point of view and his likely reaction.

Most letters are written on one page (Tables 3 and 4; see also Table 5). In a few words you must pass on your message and create the right atmosphere between yourself and the person(s) addressed. The tone of your letter will depend upon its purpose.

Table 2. Different kinds of letter and their tone

Purpose of letter	Tone
Request for details (of an appointment, a research grant, an item of equipment). *Invitation* to a speaker.	Clear, simple, direct and courteous.
Application for an appointment, a research award, etc., includes evidence of suitability and is usually supported by additional information on separate sheets (for example: details of applicant and/or details of the proposed research project).	Clear, direct and factual. Confident, but not aggressive.
Complaint	Clear and direct but not aggressive.
Reply (to an enquiry or complaint) giving information, instruction or explanation. Reply to all the points raised in the enquiry.	Clear, direct, informative, polite, helpful and sincere.
Acknowledgement (of an enquiry or application). *Acknowledgement* by postcard.	Simple and direct. Discreet.
Letter of thanks	Appreciative.

Table 3. The form of a personal letter
The address need not be punctuated. No words should be abbreviated.
The date should be given in full, without punctuation. The salutation
includes the name of the recipient, and the complimentary close is
normally: Yours sincerely. Your signature should be legible. The
name and address of the recipient is as written on the envelope.
This type of letter is used in business when the correspondents have
met or when they know one another from conversations on the
telephone or from previous correspondence.

Address of sender
(
spaces   (
(
Date
)
)

Salutation,
)
1   Information required
(
2   Supporting details
(
3   Conclusion and/or action required
(
Complimentary close
(
Signature
(
Typed name of sender.
)
)
Name and address
of recipient
)
)
)
Reference line: initials of person signing the letter
and those of the typist

Table 4. The form of a business letter

Addresses should not be punctuated. Words such as company (Co)
and limited (Ltd) may be abbreviated. The date should be given in
full, without punctuation. The name and address of the recipient
must be as written on the envelope. The salutation should be
Dear Sir, or Dear Sirs, or Dear Madam, and the complimentary
close: Yours faithfully *or* Yours truly. The supporting details,
if they are more than a few lines, should be sent on a separate
sheet which should have a title. This and any other enclosures must
be listed after the name of the sender under the heading:
Enclosures.

                               Address of sender
                                     (
                         spaces (
                                     (
                                     Date
)
)

Position and
address of
recipient
)
)
)
Salutation,
)
                      Subject heading
     1   Information required
   (
     2   Supporting details
   (
     3   Conclusion and/or action required
   (
        Complimentary close
   (
        Signature
   (
        Typed name and
        position of sender.
)
)
Enclosures:
        List
)
)
)
Reference line: initials of person signing the letter
              and those of the typist

The following memorandum on *Brevity*, written by Winston Churchill in 1940 to the heads of all government departments, is Crown copyright. It is reproduced with the permission of the Cabinet Office.

To do our work, we all have to read a mass of papers. Nearly all of them are far too long. This wastes time, while energy has to be spent in looking for the essential points.

I ask my colleagues and their staffs to see to it that their Reports are shorter.

(i) The aim should be Reports which set out the main points in a series of short, crisp paragraphs.

(ii) If a Report relies on detailed analysis of some complicated factors, or on statistics, these should be set out in an Appendix.

(iii) Often the occasion is best met by submitting not a full-dress Report, but an *Aide-memoire* consisting of headings only, which can be expanded orally if needed.

(iv) Let us have an end of such phrases as these: 'It is also of importance to bear in mind the following considerations...', or 'Consideration should be given to the possibility of carrying into effect...'. Most of these woolly phrases are mere padding, which can be left out altogether, or replaced by a single word.

Let us not shrink from using the short expressive phrase, even if it is conversational.

Reports drawn up on the lines I propose may at first seem rough as compared with the flat surface of officialese jargon. But the saving in time will be great, while the discipline of setting out the real points concisely will prove an aid to clearer thinking.

Within any organization memoranda take the place of letters. They represent the sender (and also his department). Memoranda need not be impersonal but they should be direct; giving information, suggestions or recommendations, or indicating clearly the information or action required. The paragraphs of a memorandum should be numbered (or sub-headings should be used). Numbers

Table 5. The use of a postcard
Neither the salutation nor the complimentary close should be included.

```
+-------------------------------------------------------------+
|                                                             |
|                                                    Date     |
|                                                             |
|                    *                                        |
|         Message                                             |
|                                                             |
|                                                             |
|                                                             |
|         Signature                                           |
|                                                             |
|         Typed or printed name and address of sender         |
|                                                             |
|                                                             |
|                                                             |
+-------------------------------------------------------------+
```

*Note that a postcard may be seen by other people, as well as by the person addressed. Therefore, when a postcard is used to acknowledge the receipt of a letter, the reply should not make public the contents or purpose of the letter: it should include only the date of the letter and its reference number/letters.

(1) direct the attention of the writer and reader to each important point; (2) make the writer think about what is to be said; (3) make the writer arrange the points in an appropriate order; and (4) help the reader. Each memorandum should be carefully composed; it should be as short as possible but as long as is necessary. Consider carefully to whom the memorandum should be addressed. Do not send copies unnecessarily to people who do not require them; because this wastes paper, wastes the reader's time, and indicates your lack of judgement.

Letters and memoranda should be on unlined paper and, if possible, they should be typed. A carbon copy should be kept as a record. The memoranda used for routine reporting should be

Awaiting the favour of your reply,
I remain.

Fig. 2. *All correspondence should be in standard current English (or standard American); not in the outmoded language still sometimes used in commerce (commercial jargon).*

standard report forms. These save time by helping the writer to know what is required and who requires the information; and the recipient knows where to look for any detail.

To keep each communication short and to the point any necessary supporting details or further information should be referred to briefly but sent as an enclosure. The initiator of any correspondence should state the purpose of the letter either by a clear, precise and specific heading, or in the first sentence. The reply, and any further correspondence, should begin: 'Thank you for your letter of ... about...' From these beginnings both the writer and the recipient know immediately what each communication is about (see also Fig. 2).

Letters and memoranda have not been superseded by other means of communication. Anything agreed on the telephone must be confirmed in writing. Misunderstandings are possible unless both parties have a record of their conversation. Every letter or memorandum must fit into the filing system (records) of both the sender and the recipient. It should therefore deal with one subject only. If you have to write about more than one subject, to the same person, each subject should be dealt with in a separate communication, even if these are enclosed in the same envelope.

Communication as part of science

The scientific method
Scientific research begins with a problem which may come from personal observation or from a consideration of work done by others. Problems are tackled by the method of investigation, in an attempt to obtain evidence related to a hypothesis. If the problem is stated as a question, then each hypothesis is a possible answer to the question or a possible explanation. The observations and measurements recorded during an investigation are data, and these are organized, classified and considered, and compared with other data arising from other investigations. This analysis leads to the bringing together of information from different sources, to synthesis, to the recognition of order (to classification), and to the making of generalizations (stated as norms, concepts, principles, theories and laws).

As one hypothesis is supported by new evidence and others are rejected, additional hypotheses may provide other possible explanations. Each hypothesis can be retained only for as long as it provides a satisfactory explanation for the observations accumulated on the subject. When a hypothesis is generally accepted by scientists working in the field it may be called a theory, and may lead to the statement of a principle or law which has value not only because it accounts for observations which have been made but also because it allows the prediction of what will happen in future observations and experiments.

Communication is involved at all stages in the application of the scientific method. The hypothesis upon which each investigation is based may come from the scientist's own observations but he should know of the observations and experiments of other scientists who are working on the same problem or in the same area of study. This helps to prevent unnecessary duplication of effort (but see p. 120), and should also result in a contribution to knowledge by ensuring that new observations are related to what is already known.

Hypotheses, theories and laws must be modified or discarded if at any time they are found wanting, or if a better explanation is suggested for the accumulated observations on the subject. While scientists may work alone, therefore, the scientific method makes science a co-operative venture and no work is complete until a report has been written.

The publication of research

The literature of science, a permanent record of the communication between scientists, is also the history of science: a record of the search for truth, of observations and opinions, of hypotheses that have been ignored or have been found wanting or have withstood the test of further observation and experiment. Science is a continuing endeavour in which the end of one investigation may be the starting point for another. *Scientists must write*, therefore, so that their discoveries may be known to others.

> Their purpose is, in short, to make faithful *Records*, of all the Works of *Nature*, or *Art*, which can come within their reach: that so the present Age, and posterity, may be able to put a mark on the Errors, which have been strengthened by long prescription: to restore the Truths, that have lain neglected: to push on those, which are already known, to more various uses: and to make the way more passable, to what remains unreveal'd. This is the compass of their Design.
>
> *History of The Royal Society*, Thomas Sprat (1667)

The popularization of science

Scientists must write formal accounts of their work for publication in journals which are read only by specialists, but which are accessible to scientists everywhere. Yet science is shaping our world, and whether they are pursuing knowledge for its own sake, or trying to solve practical problems, scientists must also write articles, reviews and books – about what they are doing and why, and about what other scientists are doing – for scientists working in other fields, for students of science, and for other interested people.

If we do not trouble to tell other people about science, or to discuss the impact of science on society, we should not be surprised if science and technology remain a closed book to many educated people, if the scientist is distrusted, if people do not appreciate the interdependence of pure and applied science, or if people expect too much of science.

In textbooks, scientists present science not only to tomorrow's scientists but also to those who will work and take decisions in other fields. The writer of textbooks, therefore, has a unique opportunity to interest and inform. If people do not take an interest in science while they are young they are unlikely to do so later. Writing good books for young people is one of the most important duties that each generation of scientists must perform. The younger the age-group the scientist writes for the more important is his work, because young children are quick to decide which subjects are of interest and which are not. If a child does not understand or is not interested by his first book on any subject the opportunity to interest him may have been lost.

Exercises in communication

Writing a letter

Writing a letter of application for an appointment is an exercise which most students will find useful and interesting. This is a good place to start teaching the essentials of clear, concise and courteous writing.

Preparing instructions

In science and technology, as in everyday life, we use instructions: how to assemble equipment, how to prepare a mixture, how to find a book in a library, etc. The preparation of clear instructions is a good test in the art of communication. The following exercise may be completed by anyone working alone or it may be used by a teacher of scientific writing first as set work and later in a class discussion.

Write a set of instructions headed: *How to write instructions.*

For an answer see p. 33.

4

How scientists should write

Your writing should reflect the way you think and work, and should therefore be in accordance with the requirements of the scientific method.

Explanation
Consider first the needs of the reader. What does he know already and what further explanation is required? Your purpose in scientific writing is to explain. What is it? What does it look like? How does it work? Why is it used? What have you done? Why and how was it done? What have you found?

Clarity
The clear thinking that is necessary for the application of the scientific method (in the statement of the problem, in formulating the hypothesis upon which the work is based, in planning the work and in its execution) should be reflected in the clarity of your writing and illustrations (Fig. 3).

Completeness
The treatment should be comprehensive. Every statement should be complete. Every line of argument should be followed through to a logical conclusion. Your writing should be free from errors of omission but you should show an awareness of the limitations of your knowledge.

Impartiality
Make clear any assumptions underlying your arguments, for if these

are incorrect your conclusions may also be incorrect. Indicate how, when and where your data were obtained, and specify the limitations of your work, the sources of error and probable errors in the data, and the range of validity of the conclusions (UNESCO, 1968). Show an awareness of all sides of a question. Try not to be biased by preconceived ideas, and take care not to over-estimate the importance of your work. Neither omit evidence that is against your hypothesis, nor undervalue the findings of other scientists when these seem to contradict your own.

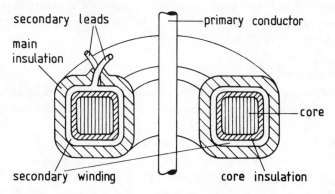

Fig. 3. *Illustrations contribute to clarity: a current transformer cut open to display the core, core insulation, secondary winding, main insulation and secondary leads.*
After Wright, A. (1968) *Current Transformers*, Chapman & Hall Ltd, London.

Any assumption, extrapolation, or generalization, should be based on sufficient evidence, and should be in accordance with *all* that is known on the subject. Any assumptions, conjectures, and possibilities discussed, should not be referred to later as if they were facts. Words to watch, because they may introduce an assumption are: *obviously*, *surely*, and *of course* (see also Table 6).

Order
The reader will find your message easier to understand if information and ideas are presented in a logical order. The need for

sufficient explanation, for clarity and completeness, and for an orderly presentation of information, is most obvious in giving instructions (p. 33 and Fig. 7, p. 87).

Accuracy
The scientific method depends upon care in observation, precision in measurement, care in recording these observations and measurements, and care in their analysis. Every experiment should be repeatable and every conclusion should be verifiable.

Apart from your good intentions; accuracy and clarity depend upon the care you take in the choice and use of words (see Chapters 6 and 7).

Objectivity
Most people respect authority, and one problem faced by anyone who has something new to say is that people are reluctant to accept

Table 6. Phrases that scientists should not use

Introductory phrases	A possible interpretation
As is well known	I think
It is evident that	I think
It is perhaps true to say	I do not know what to think
It is generally agreed that	Some people think
All reasonable men think	I believe
For obvious reasons	I have no evidence
There is no doubt that	I am convinced
It is likely that	I have not got enough evidence
As you know	You probably do not know
As you know	This is superfluous
As mentioned earlier	This is superfluous
Tentative conclusions	Possibilities
So far as we know	We could be wrong
It is not necessary to stress the fact	I should not need to tell you
The most typical example	The example that best suits my purpose

anything that conflicts with existing beliefs. In science, every statement should be based on evidence and not on unsupported opinion. Speculation cannot take the place of evidence. The scientist should therefore avoid excessive qualification. Words and phrases such as *possible, probable, perhaps, it is likely to,* and *is better referred to perhaps,* should cause you to think again. Have I considered the evidence sufficiently? Is there enough evidence for the qualification to be omitted? If not, are further investigations needed before this work is ready for publication? *The latter possibility seems quite possible.*

When no more information is available on any point, the need for further work may be mentioned. Reeder (1925) emphasized that scientists should not reason from lack of evidence against a hypothesis, or state an opinion as a fact. Do not rely on authority but on evidence. Never state the opinions of others as though these are facts; and never state the opinion of the majority as a fact.

In scientific writing nothing should be implied or left to the reader's imagination. The novelist, journalist or advertizer, to drive home a point, may repeat, exaggerate or understate his case. None of these techniques is available to the scientist who must tread a more difficult road and convince readers by evidence, relying on the truth clearly stated and on the logic of the argument.

Teleological expressions

Scientists should not endow inanimate things or even living organisms other than people, with human attributes. They should not, in scientific writing, use personal pronouns when referring to animals other than people.

Scientists should not write that *the results suggest,* nor that *another possibility suggests itself,* nor that *an experiment suggests,* since these things cannot suggest. They should not write that *the data pointed to* (since data do not point) or *from the point of view of numbers* (since numbers do not have a point of view) or *from the standpoint of soils* (since soils do not stand).

The man in the street may say that his car *does not like* a steep gradient or that the sun is *trying* to break through the clouds, but

the scientist should not allow such expressions of human emotion
to *creep* into his writing.

Simplicity
In choosing between hypotheses, the scientist is asked to prefer
the simplest explanation that is in accordance with all the evidence.
This basis for choice (that entities must not be unnecessarily multi-
plied) was first suggested by William of Occam, a theologian, in
the fourteenth century, and is known as Occam's razor.

Simplicity in writing (and in illustrations: Fig. 4), as in a
mathematical proof, is the outward sign of clarity of thought.
Scientists should write direct, straightforward prose, free from
jargon, verbosity and other distracting elaborations.

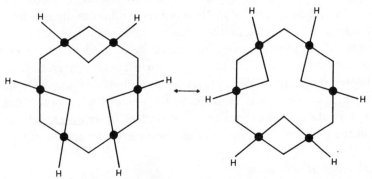

Fig. 4. *The use of a diagram to convey a new idea. The structural
formula of benzene proposed by Kekulé (1872) as a resonance
hybrid of two contributing forms.*
From Rouvray, D. H. (1975) *Endeavour* **34** (121), p. 32.

Scientific writing
Napley (1975) in *The Technique of Persuasion* advises those advo-
cates who would best serve their clients to present their case in
order, with integrity, clarity, simplicity, brevity, interest, and with
no trace of pomposity.

Explanation, clarity, completeness, impartiality, order, accuracy,
objectivity and simplicity are given here as basic requirements in

scientific writing. The writing of considerate authors has other characteristics.

Appropriateness (to the subject, to the reader and to the occasion).
Balance (showing an awareness of all sides of a question; maintaining a sense of proportion).
Brevity (the use of no more words than are needed to convey each thought to the reader; and the omission of unnecessary detail).
Consistency (in the use of numbers, names, abbreviations, symbols; in spelling and punctuation; in the use of terms).
Control (paying careful attention to arrangement, presentation and timing so as to affect the reader in a chosen way; to organization).
Interest (holding the reader's attention).
Persuasiveness (convincing the reader by evidence *forcefully* presented).
Precision (exact definition supported, as appropriate, by counting or by accurate measurement).
Sincerity (the quality of frankness, honesty, humility).
Unity (the quality of wholeness, coherence).

The preparation of a set of instructions, using words alone or words supported by effective illustrations (or samples) provides a good introduction to the essentials of scientific writing.

How to write instructions

The instructions must be *complete* so that they *explain* the action required and answer all relevant questions. They must be *clear*, *concise*, *simple*, and easy to understand. They must therefore be written by someone who has experience of the task.

1 Consider *who the instructions are for.*
2 Precede the instructions by any necessary explanation (especially the reasons for changing an established procedure).
3 List any materials required.
4 State any safety precautions; and if necessary repeat these

immediately before the step in which the precautions are to be taken.

5 Arrange the instructions in the *order* in which things are to be done.

6 Indicate the action required at each step by a separate statement.

7 Use complete sentences written in the imperative (as in this list).

8 Number the successive steps, so that the action required at each step stands out.

9 If drawings, photographs or samples are used, place each one next to the words it augments.

10 Work through the instructions.

11 Arrange for a trial by at least two other people, one with experience of the task and one without.

12 *Revise* your draft, if necessary, after the trial. Add your initials and the date.

Unscientific writing

Example 1
The complaint of examiners that students cannot write good English applies, I think, mainly to science students.... As their abilities lie outside literature, it is not surprising that science students write badly.

> *Some faults*
> 1 An opinion is expressed and later stated as a fact.
> 2 The author gives no evidence in support of the implication that students are good at either literature or science.

Example 2
Under present day conditions there can be little doubt that nitrogen is perhaps the most important factor in feeding the world. It is not necessary to stress the fact that...

Some faults
1 There is excessive qualification in the first sentence.
2 In the second sentence the writer is about to stress something which does not need to be stressed.
3 If something is stated as a fact it is not necessary to call it a fact.

Example 3
The last ten years have seen changes in teaching of a magnitude unequalled in any previous period of our educational history. Such advances have necessitated a monumental expenditure of money and human resources, and it is interesting to note that whereas in countries like the United States...

Some faults
1 Years have seen is teleological. Years cannot see.
2 Of a magnitude unequalled *means* unequalled.
3 In any previous period of our educational history is tautological; *it should read* in our educational history.
4 Changes are later referred to as advances.
5 Advances do not necessitate.
6 Expenditure cannot be monumental.
7 The words *it is interesting to note that* can be omitted without altering the meaning of the sentence.
8 Are any countries, other than the United States, like the United States?
9 The first sentence refers to education in Britain between 1964 and 1974. Is this statement true?

Example 4
Safe and efficient driving is a matter of living up to the psychological laws of locomotion in a spatial field. The driver's field of safe travel and his minimum stopping zone must accord with the objective possibilities; and a ratio greater than unity must be maintained between them. This is the basic principle. High speed, slippery road, night driving, sharp curves, heavy traffic and the

like are dangerous, when they are, because they lower the field zone ratio.

> *Some faults*
> 1 The writer's meaning is not clear. Does he mean that a driver should always be able to stop within the distance that he can see to be clear?
> 2 The writer seems to have tried to make a simple subject unnecessarily complex.

Example 5

Much of the Romagna of Italy, for instance, which was fully populated in ancient times, was only restored to its ancient population and productivity by great efforts in the present century.

> *Some faults*
>
> | | |
> |---|---|
> | 1 fully populated | How many? |
> | 2 in ancient times | When? |
> | 3 only restored ... by | restored ... only by |
> | 4 to its ancient population | Very old people! |
> | 5 and productivity | As productive as in ancient times? |

5

Think – plan – write – revise

If you wish to improve your writing, you have taken the first step by recognizing the possibility of improvement. To ensure further improvement and so provide encouragement, you should treat every composition – however small – in the same way: always think, plan, write and then revise.

Thinking and planning

The first two steps – thinking and planning – will help you to get started and take you well on the way to completing your work. You will also find writing easier if you do not expect your first draft to be perfect and are prepared to spend time on revision.

Collecting information and ideas

A common fault in writing is to spend too much time on the early part of a composition. As a result, some ideas receive too much attention simply because they are considered first and other things too little attention because they come later.

Planning may take a few minutes (as in an examination) or much more time may be spent on the search for information and ideas, and in discussion and thought. Irrespective of the time available, the first step is to organize your thoughts. A good title should help you to define the purpose and scope of the composition; and it should inform the reader. Similarly, for a commissioned report, the terms of reference must make clear what is needed, by whom, and when.

Consider your readers and anticipate their questions. They want relevant information, well organized and clearly presented, and with sufficient explanation. In conversation they would ask such questions as: Who? What? Where? When? Why? How? Ask yourself these questions. They serve as mental tin-openers (Warner, 1915). Your answers will lead you to further questions.

You know far more about many subjects than you at first suppose. In a few moments of thought and reflection you can usually make a succession of relevant notes. You may use some of these as the topics for separate paragraphs and others as supporting ideas within a paragraph. You may leave out other points, in your final selection of material, either because they provide unnecessary detail or because you select better examples.

The topic outline

As information and ideas are assembled; spread key words, phrases and sentences over a whole page (or write them on separate index cards). Use the main points as headings and note supporting details below the relevant heading. Then number the headings (or place the cards in order) as you decide:

What is the purpose and scope of the composition?
How is the subject to be introduced?
What is the topic for each paragraph?
What information and ideas must be included in each paragraph?
What diagrams are needed and where should they be placed?
What can be left out?
What needs most emphasis?
How can the paragraphs be best arranged in a logical sequence?
Would sub-headings help your readers?
How can the composition be concluded effectively?

Your topic outline contributes to order and to the organization which is essential to writing. It will help you to deal with each aspect fully in one place, to avoid digression, and to maintain the momentum which makes a composition hold together. If possible, put your topic outline on one side for a while. This will save time because it is easier to add new topics to a list, or to change the numbering of the topics, than to have second thoughts when your first draft is complete.

When you have collected all the information you require and are satisfied with your topic outline, you are ready to write. Working from the topic outline you can write with the whole composition in mind with each word contributing to the sentence, each sentence to the paragraph and each paragraph to the composition, and with meaning as the thread which runs through the whole. Only by preparing a plan can you maintain control, so that you present your subject simply, forcefully and with an economy of expression.

Table 7. Introductory and connecting phrases which can usually be deleted without altering the meaning of the sentence

---

It is considered, in this connection, that ...
From this point of view, it is relevant to mention that ...
In regard to ..., when we consider ..., it is apparent that ...
As far as ... is concerned, it may be noted that ...
It is appreciated that ... in considering ...
It is of interest to note that ... of course ...
In order to keep the problem in perspective we would like
    to emphasize that ... there is no doubt that ... not least of
    these ...
In conclusion, in relation to ..., it was found that ...
From this information it can be seen that ... in so far as ...
It is known from an actual investigation that ... as follows:
This report is a summary of the results of an enquiry into ... which,
    as you may remember, ... with respect to ...
It has been established that, essentially, ... in the case of
The evidence presented in this report supports the view that
    ... in the field of ... for your information ... in actual fact ... with
    reference to ... in the last analysis

Because the first words and last words in a paragraph attract most attention, never begin a paragraph with unimportant words. Omit superfluous phrases such as: *First let us consider.... Secondly it must be said that.... An interesting example which should be mentioned in this context is.... Next it must be noted that.... In conclusion....* Remember that the plan is for you (as an aid to thinking). It is not for the reader (who requires only the results of your thought). Superfluous introductory and connecting phrases (Table 7) distract the reader's attention. The change of subject is clearly sign-posted by the break between paragraphs and the new topic should be introduced directly and forcefully in the first words of the new paragraph.

The order of paragraphs

After the title and the introductory paragraph, further paragraphs should be arranged so that they lead logically to the closing paragraph. The logical order may be, for example, chronological or geographical. In a short work it may be an order of increasing importance, or in a long work an order of decreasing importance, or the order may be dictated by house rules or by the requirements of a customer or supervisor.

The first paragraph is your readers' first taste of what is to come. Here you must capture their interest. Your first paragraph must leave no doubt about the purpose and scope of the composition, but there are many ways of beginning (see p. 81).

There should be one paragraph for each aspect of the subject (for each topic). Each paragraph should therefore be well ordered and clearly relevant, with a limited and well-defined purpose. The topic for each paragraph is usually clearly stated (or is apparent) in the first sentence, but in an explanation or argument it may come last. All sentences in the paragraph must provide information or ideas relevant to the topic – but nothing irrelevant – and the first and last sentences should also help to link the paragraphs so that the reader can see clearly how one paragraph leads logically to the next.

Within the paragraph each sentence should convey one thought and punctuation marks should be used when they are needed to clarify meaning or to make for easy reading (Gowers, 1973; Partridge, 1953). Each sentence should be obviously related to the preceding sentence and to the next. No new statement should be introduced abruptly and without warning. The sentences within each paragraph should therefore be in a logical and effective order so that they hold together and convey your meaning precisely.

Balance is important in writing as in most things. The sentences in a paragraph and the paragraphs in an essay, like the handle and the blade of a knife, must be balanced in themselves and in relation to one another. Your composition as a whole must be well balanced: ideas of comparable importance must be given similar emphasis.

Paragraphing breaks up the page of writing, provides a pause at appropriate points in your narrative, and helps the reader to know that one thing has been said and that it is time to think of the next. Short paragraphs are also the easiest to read and so they make for efficient communication. However, paragraphs are units of thought, each with one thought or with several closely

The end should be strong, forceful, convincing and final.

Fig. 5. *The end of any composition should be a rounding off. It should leave the reader with a lasting impression of the work.*

connected thoughts, and they will therefore vary in length.

The topics covered should lead to some conclusion and/or to recommendations, or should provide the basis for speculation, or should emphasize some aspect of the subject which serves as a link between paragraphs and leads to some theory or hypothesis. Whatever method you adopt for bringing your composition to a close, the end should be obvious to the reader (Fig. 5). It should not be necessary to begin the closing paragraph, as do many inexperienced writers, with the words: In conclusion. . . .

> The writer I like has paragraphs varied in length, development and organization. He . . . moves quickly through simple material, and explains . . . difficult points. His paragraphs are carefully connected, and when there is a marked change in thought, there are enough indications to help me follow the shift. He does not repeat unnecessarily or digress; instead he covers his subject thoroughly and briefly. While I am still interested, he completes his work in a satisfactory final paragraph. . . .
>
> *Effective Writing*,   H. J. Tichy (1966)

Writing

With the plan complete, the theme chosen, and the end in sight, try to write your composition at one sitting and use the words that first come to mind. Stopping for conversation, or to revise sentences already written, or to check the spelling of a word, or to search for a better word, may interrupt the flow of ideas and so destroy the spontaneity which gives freshness, interest and unity to your writing. The time for revision is when the first draft is complete.

Work from your topic outline. Present detailed knowledge in a logical, interesting and objective way. Use enough words to make your meaning clear. Too few words will provide insufficient explanation and too many may obscure meaning and will waste the reader's time. Avoid figurative language because this may confuse some of your readers.

The use of the imagination may be encouraged in an English essay, and there is a place for imagination and speculation in science (in the formulation of hypotheses) but non-fiction should be impartial, accurate and objective. When the interpretation and assessment of evidence calls for the expression of an opinion this must be clearly stated as such.

Arguments in favour of any idea expressed should be based on the evidence summarized in your composition, and all your statements should be supported by examples, so that the reader can judge their validity. Sources of information and ideas should be indicated in the text (for example, by the author's name and the date of the publication referred to). Criticism of other people's work must be reasoned and not based on preconceived ideas for which there is no evidence.

Revising

Two processes are involved in written communication. The first, in your mind, is the selection of words to express your thoughts. The second, in the mind of the reader, is the conversion of the written words into thoughts. The essential difficulty is in trying to ensure that the thoughts created in the mind of the reader are the same thoughts that were in your mind.

> The only proper attitude is to look upon a successful interpretation, a correct understanding, as a triumph against the odds. We must cease to regard a misinterpretation as a mere unlucky accident. We must treat it as the normal and probable event.
> *Practical Criticism*,   Professor I. A. Richards (1929)

Too often the reader, looking at an ambiguous sentence or at a sentence that is obviously incorrect, must try to work out what the writer meant (Table 14, p. 90). If you take pride in your work you must revise carefully to try to ensure that your words do record your thoughts. Try to ensure that the reader takes this same meaning.

A common failing in writing is to include things in one place which should be in another. Indeed, one of the most difficult tasks is to get everything into the most effective order. One reason for this, even after careful planning, is that we think of things as we write. Things are then included in one paragraph although they could be better placed in another, or even under another heading.

In writing we use words as they come to mind but our first thoughts are not necessarily the best and they may not be arranged in the most effective order. Wrong words and words out of place lead to ambiguity and distract the reader's attention, and so have less impact than would the right words in the right place. By further thought, intelligent people should be able to improve their first draft.

Revise carefully so that readers do not have to waste time on an uncorrected first draft which may reflect neither your intentions nor your ability. Read the whole composition aloud to ensure that it sounds well, and that you have not written words or clumsy expressions that you would not use in speech. (See Table 8, p. 48; and the check lists on p. 84 and 147.)

Thinking, planning, writing and revising are not separate processes, since writing is an aid to thinking. The time taken in planning, writing and revising is time for thought. It is time well spent, for when the work is complete your understanding of the subject will have been improved.

To admit that you need to plan your work, that your first draft is not perfect, that you need to revise your first drafts, and that you can benefit from the comments of a colleague or from the advice of an editor, is not to say that you are unintelligent. Even after several revisions you may not appreciate all the difficulties of the reader. It is a good idea, therefore, to ask at least two other people to read your corrected draft of an important communication. Preferably one reader should be an expert on the subject and the other should not be. Coming fresh to the work they will see things that are not sufficiently explained, that are irrelevant (not necessary or out of place) and which are ambiguous or do not convey the thought that you intended.

Because the quality of your writing reflects upon your employer as well as upon yourself, some employers have a procedure for editing and revising manuscripts. Your employer may also wish to ensure that nothing confidential or classified as secret is reported. You should also remember that in talking or writing about your work you may invalidate a later patent application. If you need advice, consult a patent agent.

The function of a critic is to improve your writing and any comments should be welcomed and should be considered when you revise your work. Because of this, do not ask people to read a draft unless you respect their judgement and can rely on them to give an honest opinion. You are fortunate if you know someone who will criticize your writing as well as the subject matter.

The readers will see mistakes, ambiguities, badly presented arguments, and superfluous words and sentences, which are immediately obvious when someone points them out. They will see good points which require more emphasis (p. 82). Experienced writers learn to improve their first drafts but they can still benefit from a reader's frank comments.

When a paper has been revised it is a good idea to put it on one side for a while. One way to do this is to send a copy to another reader. There is bound to be a delay and this is the time to get on with something else before you reconsider the composition. The value of a second reader's opinions is that they may reinforce the views of the first reader, and new comments may be made.

An important article or report will probably be typed several times. Every time it is put on one side and then reconsidered, and every time it is read by someone else, further improvements will be made. Each draft should be easier to read, easier to understand, and therefore more interesting than earlier drafts.

To see how hard writing is, even for experienced writers, we have only to study their manuscripts; they are full of alterations, crossings out, additions, loops, arrows, blots. And what is hard for them is also hard for us (Vallins, 1964). The apparent spontaneity of easy-reading prose is the result of hard work; for every writer needs to correct and improve his first drafts.

On the whole, I think the pains which my father took over the literary part of the work was very remarkable. He often laughed

or grumbled at himself for the difficulty which he found in
writing English, saying, for instance, that if a bad arrangement
of a sentence was possible, he would be sure to adopt it ...
When a sentence got hopelessly involved, he would ask himself
'now what do you want to say?' and his answer written down,
would often disentangle the confusion.

The Life of Charles Darwin,   Francis Darwin (1925)

Ernest Hemingway rewrote the last page of *Farewell to Arms* 39
times before he was satisfied with it. Aldous Huxley said: 'All my
thoughts are second thoughts'. H. G. Wells wrote a first draft
'full of gaps' and then made changes between the lines and in
the margin. He revised the whole work as many as seven times
(Tichy, 1966).

Those who write best probably spend the most time criticizing
and revising their prose; making it clear and concise but not stulti-
fied; and ensuring a logical flow of ideas. However, writing is only
one part of a scientist's work and there comes a time when the
task of revision has to stop. Furthermore, revision must not be
taken so far that the natural flow of words is lost, for language
that is artificial in its bluntness and simplicity may lack interest
and style. Alan Sillitoe said of *Saturday Night and Sunday Morning*:
'It had been turned down by several publishers but I had written
it eight times, polished it, and could only spoil it by touching it
again'.

The pleasure to be derived from writing comes from the effort
of creative activity – which may lead you to a deeper understanding
of your subject. Each composition is original: it is a vehicle of
self-expression, a presentation of information and ideas in a way
which is peculiar to the writer. No two people will select the same
material for inclusion, arrange the arguments in the same way,
make the same criticism, or reach the same conclusions.

Pleasure comes from writing something which will affect other
people. The reader may be persuaded or convinced by evidence
logically presented, or may be annoyed or misled by poor writing.
Each communication is a challenge to the writer to present infor-

mation and ideas directly and forcefully, to help the reader along, and to affect the reader in a chosen way; for this is the purpose of all exposition.

Practise essay writing

Your first practice in the art of composition comes from writing essays. Some teachers of English (but not all; see Graham-Campbell, 1953) approve of a different style from that required in science. In an English essay the approach does not have to be systematic; the theme may be developed without formal argument, and imaginative writing is encouraged. The writer may not always strive for clarity: the reader may be left to pull out the plum.

To a scientist, an essay is a means of conveying information and ideas. It is a short written account of a well defined subject. It is clear and decisive, systematic and comprehensive.

Practice in writing will help you to think and to organize your thoughts in an essay, or to write instructions, or an account of an observation or experiment, or a summary of someone else's work, in English that can be understood by the reader. Every teacher of science should therefore set exercises in writing and should be able to give encouragement and constructive criticism.

However, an essay is not only an exercise in thinking and writing for students but also a vehicle in which any writer's thoughts are assembled and organized (as in a magazine article or review) and conveyed to the reader in a clear, concise and interesting way. It should follow from this that, before you write, you should decide for whom you are writing, and you should not write unless the subject is one in which you are interested and about which you have something interesting to say (Fig. 18, p. 316).

On the question of the kind of subject the young writer should be expected to tackle, Henn (1960) writes:

> I have little faith in the practice of inviting young writers to produce long essays on abstract subjects, though I am aware that this exercise is fashionable at many stages of education ... It seems likely that general practice in writing should be based

Table 8. How to write

Think	1	Consider the title or the terms of reference.
	2	Define the purpose and scope of the composition.
	3	Consider the time available and allocate this time to thinking, planning, writing, and revising.
	4	Make notes of relevant information and ideas.
	5	Decide what the reader needs to know. If possible, identify your readers and prepare a circulation list.
Plan	6	Prepare a topic outline.
	7	Underline the points which require most emphasis.
	8	Decide upon an effective beginning.
	9	Number the topics in a logical order.
	10	Decide upon an effective ending.
	11	For a report, decide what help you will need with the preparation of illustrations, photography, copying, etc.
Write	12	Write on wide-lined A4 paper with a 25 mm margin.
	13	If possible, put other tasks on one side and see that you are free from interruption.
	14	Use the topic outline as your guide.
	15	Use effective headings and keep to the point.
	16	Start writing and keep going until you have completed your first draft, using the first words that come to mind.
Revise	17	Does the composition read well; is it well balanced?
	18	Are the most important points sufficiently emphasized, and is anything essential missing?
	19	Are there faults of logic or mistakes in spelling?
	20	Is the meaning of each sentence clear and correct?
	21	Are any long sentences well organized?
	22	Does the writing match the needs of your reader(s) in style, vocabulary, abbreviations, symbols, mathematics, and illustrations?
	23	Put your composition on one side for a while and then revise it again.

---

upon some subject – which may, perhaps, be a hobby – that
has already excited interest and some degree of mastery.

1 When you have to write an essay choose a subject that interests
you and which you know something about, *or* read about the
subject before you start to write so that you can select, arrange,
and maintain control of your material.

2 Learn to improve your writing by studying the techniques of successful essayists (see Graham-Campbell, 1953). Consider, for example, the purpose and scope of a leading article in a good newspaper, or an article in a scientific magazine.
Does the title capture your interest?
Does the opening sentence make you want to read the article?
Reconstruct the writer's topic outline by picking out the topic for each paragraph.
Is each paragraph relevant to the title?
Are the paragraphs in a logical order?
Do they lead smoothly to an effective conclusion?
Study one paragraph. Can you understand the idea presented in each of the sentences of the paragraph? Are all the ideas relevant to the topic? Why are the ideas presented in this order? Could they logically have been arranged more clearly in any other sequence?

These suggestions (1 and 2 above) may also be used by teachers of scientific writing (as a basis for class work and/or discussions). The next three could be set by teachers of science and then discussed in class.

3 Write a topic outline for an essay on a scientific or technical subject which is of particular interest to you.
Your plan should comprise key words and phrases, the topics for paragraphs, with brief notes on each topic to remind you of information and ideas to be included in the paragraph. Use one page for rough work and then arrange your plan neatly on the next facing page.

4 Of what use is the technique of essay writing to a scientist? Write a set of instructions headed *How to write an essay* for someone who does not know what is required in an essay on a scientific subject (see Table 8).

5 Describe a laboratory technique that you have used, in the form of a set of instructions which could be followed by someone with a similar training to yourself but with no experience

of this technique. Explain the scientific basis for the technique. What are its applications?

Students will find that they learn about the subject at each stage in the preparation of an essay: they learn from gathering information and ideas, from selecting and arranging their material, from writing, from revising (as if they were correcting and marking the work) and if necessary from rewriting.

Many teachers ask for a topic outline to be handed in with each essay, and talk to their students after the work has been marked with corrections and suggestions.

If possible, examination scripts should be returned to students so that they can see their mistakes and appreciate where they have failed to express themselves clearly.

The following notes on examination techniques can be understood by students working alone, but they may also be used by teachers in class discussions both before and after a test or examination.

Examination technique

How examination papers are set and marked
Candidates should consider how examination papers are marked so that in planning their answers they can attempt to score marks. Each part of an answer, and each paragraph, should be seen as an opportunity to gain marks by adding relevant information and ideas, and by showing understanding.

1 In most examinations marks are divided equally between the questions to be answered so that there are 25 marks per question when four questions are to be answered, and 20 marks per question when five are to be answered. You must, therefore, answer the right number of questions.

2 To be fair to all candidates, examiners allocate the marks which may be obtained for each question according to a marking scheme.

This scheme is a topic outline.

If the question is set in parts, a certain number of marks will be allocated to each part of the answer.

Whether or not the question is set in parts, the examiner will expect the student to refer to, and to show his understanding of, all those things which are relevant to the answer.

A student who does not answer all parts of any question or who gives an answer which is otherwise incomplete, cannot score full marks on that question.

Making the best use of your time in an examination

1 Read the instructions at the head of the question paper.

2 Read all the questions carefully until you are sure that you know what is required in each answer. Then select the questions which you can answer most fully. Otherwise you may realize, after leaving the examination, that you could have answered another selection of questions and obtained better marks.

3 Before you plan your answer read the question again to make sure that you understand what is required. Answer the question you have been asked and not a similar question which you were hoping for.

4 For essay-type questions, plan all your answers quickly at the beginning so that you can reconsider each topic outline before you start to write.

5 Answer the required number of questions. If all questions carry the same number of marks, divide your time equally between the questions. Do not spend more time on those questions you know most about. Remember that it is easier to score half marks on a question that you do not know much about than it is to score full marks when you think you can write a good answer. The first few marks are the easiest to obtain, with a little thought, if you know anything about the subject. But a little extra time spent on a question, upon which you have already spent long enough, is likely to be less rewarding.

6 Keep an eye on the time. Allow a proportion of the time

available for reading all the questions at the beginning, for planning your answers, and for reading through your work at the end to correct any slips of the pen and to add important points that you did not remember the first time through.

7 Do not waste time. Arrive at the examination before the start. Do not waste time during the examination. Do not leave before the end.

Answering questions in an examination

1 Do not make vague statements. Give reasons and examples.

2 Do not leave things out because you think they are too simple or too obvious. Do not include anything that is irrelevant, but make sure that everything relevant is included and clearly explained, however briefly, to show your knowledge and understanding. Remember that the examiner cannot assume that you know things, and can give you marks only for what you write.

3 If you include anything that is not obviously relevant, explain why it is relevant. An examination is not simply a test of your ability to recall facts and ideas. It also provides an opportunity to show your ability to distinguish relevant from irrelevant material.

4 Make things as easy as you can for the examiner. Write clearly. Get to the point quickly and keep to the point. Plan your answer so that it is well organized and well balanced, and so that, without digression or repetition, you can say everything you wish to say in the time available.

5 Make sure that any diagram is simple so that it can be completed quickly and neatly. Use coloured lines to represent different things. Do not waste time on shading.

6 If you have read a book, a review or an original paper, relevant to your answer, refer to this by giving the name of the author (and the date of publication), to show the source of your information.

7 If a question is set in several parts, answer all the parts and answer them in the order in which they are set – because

the examiner is expecting to mark the parts in this order.

8 If a question is set in several parts, answer each part separately; and if the parts are numbered use the same numbers to indicate the separate parts of your answer. If the parts are not numbered, use appropriate headings to draw attention to the parts of your answer.

9 If you are asked to discuss, you must discuss all sides of the question and refer to any unresolved problems.

10 If you are asked to compare, you should also refer to any differences (even if the question does not say compare and contrast).

6

Thoughts into words

Vocabulary

Word games are popular because there is fun to be had from words. The habit of consulting a good dictionary, whenever you come across a word that you do not understand, can be a life-long source of enlightenment and pleasure.

Our interest and pleasure in words is not surprising since when we speak or write we are trying to put our thoughts into words. Indeed the use of words is even more fundamental; for without words we cannot think. We are limited in our ability to think by the number of words at our command. If we have a large vocabulary and can construct effective sentences and paragraphs, we are better able to express ourselves.

We write so that we can tell others what we think, but if we use words incorrectly, or use words that our readers do not understand, we shall be misunderstood. We must think about words so that we can use them correctly and choose those that we expect our readers to know.

English is the language used by most scientists for international communication, and people who read English as a second language are most likely to understand plain words in simply constructed sentences. If you wish to be widely understood, try to express your thoughts in simple language.

There are problems of vocabulary for anyone who wishes to popularize science. Even books, magazines and articles in newspapers include difficult words and technical terms: instead of bridges, barriers are built between specialists and other educated people. After discussing this problem, Flood (1957) lists 2000 words which are 'adequate for the presentation to a non-scientific reader of all in science that he could reasonably be expected to understand'. Most educated people will know these words which may be used to explain new concepts prior to the introduction of additional words.

One of the delights of English is its rich vocabulary. No two words have quite the same meaning, and the choice of one word when some other word makes more sense will not help the reader. When *The Times* reported that Rudyard Kipling was to be paid £1 a word for an article, an Oxford undergraduate sent Kipling £1 and asked, 'Please send us one of your best words'. He replied, 'Thanks'.

The right word is not always the first to come to mind, and people who have too few words at their command may fall back upon hackneyed phrases or clichés (such as: *people are more important than things*; *to leave severely alone*; and, *last but not least, to leave no stone unturned*, in *pushing back the frontiers of knowledge*). Instead, they should take trouble to find the word or words which express their meaning precisely. A rich vocabulary gives us a selection of words with which we may express our thoughts:

'My dear, a rich vocabulary is the true hallmark of every intellectual person. Here now' – she burrowed into the mess on her bedside table and brought forth another pad and pencil – 'every time I say a word, or you hear a word, that you don't understand, write it down and I'll tell you what it means. Then you can memorize it and soon you'll have a decent vocabulary. Oh, the adventure,' she cried ecstatically, 'of moulding a little new life!' She made another sweeping gesture that somehow went wrong because she knocked over the coffee-pot and I immediately wrote down six new words which Auntie Mame said to scratch out and forget about.   *Auntie Mame*, Patrick Dennis (1955)

You may use words that you understand, and that your readers understand, yet still write sentences that are difficult to read.

Flesch (1962) asks people not to fill their writing with words that have been made longer by having bits added (for example: prove, approve, disapprove, disapproving; nation, national, nationalize, nationalization, denationalization). He does not suggest that such words should never be used, but that long involved sentences with many long words make for hard reading. Prefer a short word to a long one (Table 9) unless the long word is better, and a single word to a phrase if brevity makes for clarity (see also Gowers, 1973 *Plain Words*).

Some people like fashionable words: deprived, dialogue, disadvantaged, escalation, hopefully, integrated, meaningful, obscene, overall, paradigm, relevant, traumatic. (See also idiom, p. 67.)

Table 9. Prefer a short word to a long word if the short word is more appropriate

Prefer this ...	to this	Prefer this ...	to this
do	accomplish	suggest	hypothesize
extra	additional	reputation	image
expect	anticipate	sign	indication
help	assistance	person	individual
simple	simplistic	people	individuals
use	application	please	kindly
discovery	breakthrough	methods	methodology
begin	commence	change	modification
about	concerning	partly	partially
guess	conjecture	preventive	preventative
so	consequently	about	regarding
much	considerable	is	represent
build	construct	show	reveal
show	demonstrate	shortened	streamlined
meet	encounter	later	subsequently
except	excepting	enough	sufficient
show	exhibit	end	terminate
build	fabricate	use	utilize
first	firstly	almost	virtually
send	forward	guidance	guidelines

Today, almost nothing is done with a purpose; it is *performed with an objective*, preferably *performed objectively*, and always by *personnel* (McClelland, 1943).

When people think that something is too technical, it is just as likely that the writing is at fault. Yet some writers seem to think they impress people when they use long words (p. 75). Their studied avoidance of short words is not likely to impress, and is very likely to annoy, confuse (or amuse). This anonymous version of a well-known nursery rhyme pokes fun at grandiloquence:

> Scintillate, scintillate, globule aurific,
> Fain would I fathom thy nature specific,
> Loftily poised in the ether capacious,
> Strongly resembling a gem carbonaceous.

Some writers seem to think that scholarly writing must be hard reading, and that a pompous style is necessary to demonstrate to the world that they are educated. The professorial use of pompous language is copied by coteries of like-winded students (Tichy, 1966).

The meaning of words

As a guide to the meaning of words, to their origins, and to their spelling, there should be a good dictionary in the home of every educated person and on the bookshelf of every student.

If more than one spelling is correct for any word, you should be consistent: in spelling, in the use of hyphens, and in the use of capital letters.

The habit of writing a word in inverted commas to indicate that it is not quite the right word, or that it is not used in the usual sense, or that more is implied than is said, is likely to confuse people. Instead, choose the word or words which convey your meaning precisely.

Note the meaning of the following: accept (receive); except (not including); amount (mass or volume); number (of things counted); affect (to influence); effect (to cause *or* a result); complement (to make complete); compliment (to congratulate); its (a possessive pronoun); it's (it is); majority (the greater number or part); most (nearly all); principle (law); principal (main); stationary (not moving); and stationery (writing paper); uninterested (not interested); disinterested (impartial).

The following words, like the measurements made by scientists, should contribute to precision in scientific writing:

Approximate(ly) means very close(ly) and should not be used when about or roughly would be better.

Data (L. *dare* to give) refers to things given, to facts of any kind, such as the measurements which you record as numbers. It is incorrect to speak of raw data or of real data; but it is correct to refer to your own observations as original data.

Results are obtained by the analysis of data.

Statistics are numerical data systematically collected. The name statistics also refers to the science of collecting, classifying and using statistics.

Range: a word which should not be used for things which are not at the ends of a series. A range cannot start from zero.

Significant is a statistical term with a precise meaning. Scientists should try not to use it in other contexts.

Infer does not mean the same as imply. The writer or speaker implies something but the reader or listener infers.

Often: people who eat mushrooms often die (but people who do not eat them die only once).

Refute should be used in the sense of proving falsity or error and not as a synonym for deny or repudiate.

Comprise (not comprise of).

Different from (not different to).

Superior to (not superior than).

Scientists should not write *per* for a, or *re* for about.

Many people misuse the following: alternatively (for alternately); alternatives (for more than two things); centre (for middle); centred around (for centred on); circle (for disc); degree (for extent); either (for each or both); except (for unless); generally (for usually); homogenous (for homogeneous); if (for although); improvement (for alteration); lengthy (for long); limited (for few, small, slight or narrow); major (for great); minor (for little); natural (for normal); optimistic (for hopeful); optimum (for highest); percentage (for some); provided that (for if); quite (for entirely or rather); rudimentary (for vestigial); several (for some); same (for similar); singular or unique (for rare or notable); often (for in many places); always (for everywhere); sometimes (referring to place instead of time); superior (for better than); transpire (for happen); view (for opinion); virtually (for almost); volume (for amount); weather (for climate); wastage (for waste), and while (for although); see also Gowers (1973).

The meaning of words may change so much that they lose their value, and the incorrect meaning may come into common use. The new usage may remain incorrect or it may gain acceptance but the scientist should not lead the way in giving new meanings to every-day words:

Literally: a word used incorrectly to affirm the truth of an exaggeration.

Progress means a move forward, a change from worse to better, but the word is misused for change of any kind. Indeed, the most outrageous suggestion acquires a certain respectability if someone calls it progress (Orwell, 1937).

Sophisticated was once an uncomplimentary word implying sophistry and even artfulness but it is now commonly used to mean complicated or to imply that a new instrument is in some way better than an earlier model.

Viable is a term which denotes the capacity to live, but in other contexts *not viable* may mean too expensive or will not work.

Vital means essential to life and should not be used in any other context.

Words with only one meaning should not be qualified (Table 11). *Facts*, for example, are verified past events; things observed and recorded; *data*; things known to be true. It is wrong, therefore, to refer to the *fact that* energy *may* be involved or to write that the *evidence* points to the *fact* or that someone has got his *facts wrong*, and to speak of the *actual facts* is to say the same thing twice (Table 10: *Tautology*).

Hypotheses, theories and laws are not facts but are attempts to explain facts. The scientific method depends upon the formulation and testing of hypotheses. There may be conflicting hypotheses and if one of these gains general acceptance it may come to be known as a theory. In everyday usage the words theory and

Table 10. Tautology: saying the same thing twice using different words

---

Every individual one; may possibly go; on Friday 28th November next; the reason for this is because; in actual fact; one after another in succession; in the rural countryside; as an extra added bonus; I tentatively suggest; in my own personal opinion; on pages 1–4 inclusive; that by advance planning; will disappear from sight; in equal halves; in two equal halves; continue to remain; symptoms indicative of; temporary loan; but ... however; enclosed with this letter; or alternatively; grouped together; and ... moreover; superimposed over each other; topographical features.

---

idea may be used as synonyms but the scientist should consider what meaning he wishes to convey and use the following words with care: speculation, supposition, conjecture, idea, impression, surmise, expectation, assumption, presumption, guess, opinion, view, notion, hypothesis, theory and law.

Many other words have a precise meaning in the language of science – they are technical terms – but they have additional meanings in common English usage (for example: allergy, neurotic, subliminal). In scientific writing, care must be taken to use such words in the restricted scientific sense.

Technical terms
Technical terms may be a barrier to effective communication. Yet such terms are used even when they are not needed. Sometimes the writer has not troubled to avoid them. Sometimes he does

Table 11. The unnecessary qualification of words

Incorrect	Correct
absolutely perfect	perfect
the actual number	the number
an actual investigation	an investigation
not actually true	untrue
almost unique	not unique
almost perfect	imperfect
by means of	by *or* using
a categorical denial	a denial
completely surrounded	surrounded
conclusive proof	proof
cylindrical in shape	cylindrical
deliberately chosen	chosen
an essential condition	a condition
facing up to	facing
they are in fact	they are
few in number	few
green in colour	green
a positive identification	an identification
small in size	small
streamlined in appearance	streamlined
stunted in growth	stunted
swampy in character	swampy
quite impossible	impossible
quite obvious	obvious
hard evidence	evidence
real problems	problems
realistic justification	justification
they really are	they are
really dangerous	dangerous
the smallest possible minimum	the minimum
valid information	information
very necessary	necessary
very relevant	relevant
very true	true
wholly new	new

not realize that he is using words which most educated people will not understand (technical jargon) or using words in a sense which differs from common usage.

When a writer uses technical terms he makes two assumptions, which are not always justified: that the reader is familiar with the concept, and that he recognizes the concept by its technical name (Flood, 1957). Before using a technical term, therefore, you should consider whether or not it will help your readers.

Use technical terms when they are needed, but never to impress non-technical readers. People may ignore our views, and even doubt our honesty, if we insist on concealing our thoughts (or lack of thought) behind a smoke-screen of professional jargon (Smith, 1922). Wherever possible, replace a technical term by an everyday word if this can be done without altering the meaning of the sentence. If your report is for non-scientists, or if a term is defined differently by different people, any technical term must be sufficiently explained in simple language. Help your readers by relating a new word to familiar words, by indicating the nature of the thing named, by providing a brief explanation or derivation in parenthesis, by a negative interpolation, or by explaining the concept fully before giving the name of the concept (Flood, 1957).

If a technical term is used as a substitute for an explanation, it gives no more than an impression of knowledge. For example, the behaviour of an animal may be described as instinctive but few scientists attempt to define the word *instinct* which provides a cloak for our ignorance. Other words which sound like technical terms but which cannot be defined are *libido* in psychology and *ore* in geology. Unless a technical term can be defined clearly and then used with accuracy and precision, it may conceal our ignorance and obscure the need for further research, and it should have no place in scientific writing.

It is with words that we do our reasoning, and writing is the expression of our thinking.... Words and phrases that do not have an exact meaning are to be avoided because once one has given a name to something, one immediately has the feeling that the position has been clarified, whereas often the contrary is true.

The Art of Scientific Investigation, W. I. B. Beveridge (1968)

If any technical term is to retain its value, scientists must use it correctly – in the same way as other specialists. If there is no internationally accepted definition, they should say whose definition they are following (and give this definition) or they should define the term to make clear their usage. The use of any technical term must also be consistent throughout the report.

Many technical terms play an essential part in the prose of science. If they are widely accepted they contribute to an economy of words, and they should also form part of the common language used by scientists everywhere. However, other technical terms are short-lived because they serve no useful purpose, or because of misuse, or because they are never clearly defined in an acceptable way, or because they are related to hypotheses and theories which have been superseded. If you can express yourself clearly without technical terms you will be understood more widely, and your writing may be understood better by future generations.

Definitions

We may use a word in an appropriate context and yet have difficulty in defining it precisely. This is why students are asked to define terms in examinations.

When you have to define a word, to test yourself or as a class exercise, note the points which must be included (like a topic outline) and then write your definition. Proceed from the general to the particular. That is to say, state the general class to which the thing to be defined belongs and then the features which are peculiar to the thing defined. Your definition must be as simple as possible but it must apply to all instances of the thing defined, and it may be followed by an example.

Abbreviations

The names of journals are usually abbreviated in lists of references (see p. 141). Otherwise, only essential abbreviations should be used. As with complete words they must be understood by the reader, and abbreviations which are commonly used in one country may not be understood in another. Also, one abbreviation may have several meanings and even after referring to a dictionary of abbreviations the reader may still not know which meaning the writer intended. Any essential abbreviation should be written in full when it is first used and then abbreviated in parenthesis. Authors should also be consistent in using abbreviations and in their punctuation.

Gowers (1973) advises: Do not say *a priori* when you mean *prima facie*. However, it is not necessary to use either phrase. Scientists writing in English should try to convey their meaning without using phrases from another language, and even abbreviations of such phrases. Some abbreviations, such as *loc. cit.* (in the place cited), *op. cit.* (in the work cited), and *ibid* (in the same work), like the words former and latter, may contribute to ambiguity and they should not be used: see p. 124 (ISO 215). Even the abbreviations *i.e.* (*id est*: that is); and *e.g.* (*exempli gratia*: for example), are misused and, therefore, misunderstood by some people. Write *namely* (not *viz*) and prefer *about* to *approximately*, *circa*, *ca.*, *c.*, and ∼. The abbreviation etc. (*etcetera*: and other things), used at the end of a list, conveys no additional information, except that the list is incomplete. It is better to write *for example* or *including* immediately before the list.

These examples show the use of the full stop to indicate abbreviations. However, many abbreviations are not punctuated (for example: WHO: World Health Organization) and punctuation marks should not be used with SI units (International System of Units, see p. 98). Another rule is that an s should not be added to an abbreviation (except for nos: numbers; and figs: figures in the sense of illustrations).

Nomenclature

People must name things so that they can talk and write about them. Difficulties arise because different people name the same thing in different ways (use different systems of nomenclature) or because they give more than one name to the same thing. These are reasons why scientists would like to give every chemical, every kind of rock, every species of organism, every disease, every part of the body, and every other thing, its own unambiguous and internationally accepted name. Sources of information on nomenclature in the different sciences are listed in the Royal Society's *General Notes on the Preparation of Scientific Papers for Publication*.

The use of trade names (with capital initial letters) is sometimes necessary but these may not contribute to accuracy since, for example, the chemical composition of a product or the components of an instrument may change while the trade name is unchanged.

7
Using words

In a dictionary each word is first explained and then used in appropriate contexts to make its several meanings clear. For words do not stand alone: each word gives meaning to and takes meaning from the sentence, so that there is more to the whole than might be expected from its parts.

Words in context
Kapp (1973) considers that scientists must not allow words to carry either more or less meaning than they do in common usage. However, it is not easy to do this because we cannot separate a word in science from our everyday experience. It is the function of other words in a sentence to tie each word down so that the sentence as a whole has only one meaning.

Some people have favourite words and phrases, such as: also, apparently, case, found, incidentally, in fact, make, occur, of, quite, show, and use. The use of a word twice in the same sentence, or several times in the same paragraph, or many times on the same page, may interrupt the smooth flow of language, and writers usually try to avoid such undue repetition. But scientists should not be afraid to repeat a word. The right word should not be replaced by a less apt word for the sake of elegant variation. Moreover, a word repeated may give emphasis to an important point.

The position of a word in a sentence may also reflect the emphasis you wish to put upon it. An important word may, for example, come near the beginning or near the end, and in either position it may help to link the ideas expressed in successive sentences.

The position of a word may also transform the meaning of a sentence. For example, the word *only* is well known for the trouble it may cause when it is out of place. 'The words no doubt should *only be used* if the idea of certainty is to be conveyed.' The words in italics should read: be used only.

The words: *but* and *even*; *this*, *that* and *it* (*he* or *she*) and *one*; *former* and *latter*; and *other* and *another*, must be used with care or ambiguity may result. If necessary a noun should be repeated.

Baker (1955) drew attention to the German–American imitation of English in which qualifying words are piled in front of nouns (e.g. iron-containing globules, for globules containing iron; a hyaluronidase-treated area, for an area treated with hyaluronidase). Baker wonders why scientists should avoid simple English when writing for scientific journals, for the English written by those who know and love the language can scarcely be surpassed for its clarity, directness and simplicity.

Scientists should write in standard English (or standard American) and should avoid colloquial English and slang.

Standard English: the language used by educated English people.
Colloquial English: the English used between close friends, including such contractions as don't and won't.
Slang: highly colloquial language including new words or words used in a special sense which might not be understood by educated English people.

Partridge (1965) gives, as an example of the difference between standard English (or standard American): man (standard), chap (colloquial) and bloke, cove, cully, guy, stiff, or bozo (slang).

Avoid idiomatic expressions, in which the words have a special meaning that may not be understood by foreigners, not only because they will be misunderstood by some readers but also because many expressions used by one generation may be unknown to the next. Ready-made phrases also make less impact than does something

new. They indicate that the writer has not troubled to choose words
to convey his meaning precisely.

> 'As soon as certain topics are raised, ... no one seems able to
> think of turns of speech that are not hackneyed: prose consists
> less and less of *words* chosen for the sake of their meaning, and
> more and more of *phrases* tacked together like the sections of
> a prefabricated hen-house.'
>
> *Politics and the English Language,*   George Orwell (1950)

In this way, people deny themselves the simple pleasure of putting
their own thoughts into their own words. Instead, always:

> 'Open a new window,
> Open a new door.
> Travel a new highway
> That's never been tried befors.'
>
> *Mame,*   Lyric by Jerry Herman (1966)

Superfluous words
The use of too many words is a more common fault in writing
than the use of the wrong word; and while a summarizing or
qualifying phrase may help the reader (see also comment words
and connectives, p. 77), any unnecessary words can only confuse,
distract and annoy. Also, when too many words are used, time,
paper and money are wasted in typing, printing and advertizing.
In revising, therefore, reconsider each sentence and each paragraph
to see if it is necessary and prune sentences to remove all superfluous
words (see Tables 7, 12 and 13). A well constructed sentence
should have neither too many words nor too few; each word should
be there for a purpose.

Laziness in sentence construction may cause the writer to use
jargon or to choose phrases made safe by common usage in prefer-
ence to more appropriate words. Jargon may also result from
attempts at elegant variation (p. 66). Quiller-Couch (1916) con-
demned jargon, and recommended writers to prefer transitive verbs
and use them in the active voice (we obtained the following results,
not the following results were obtained) to prefer concrete nouns

(things which you can touch and see) to abstract nouns, and to prefer the direct word to the circumlocution. He listed words to watch, for those who wish to avoid jargon: case, instance, character, nature, condition, persuasion, and degree. Other indicators of jargon are: area, angle, aspect, fact, field, level, situation, spectrum, time, and type (Tables 12 and 13). Of course there is nothing wrong with any of these words in its proper place.

Many introductory phrases and connectives can be deleted without altering the meaning of the sentence (Table 7, p. 39; see also Table 6, p. 30). The writing is then easier to read. Too many words may be used in text references to tables and figures:

Fig. 7 shows that ...
It is clear from a consideration of Fig. 7 that ...

These introductory phrases are not necessary; and they may cause the reader to think that the figure shows only one thing. It is better say whatever you wish to say about the illustration and then to refer to the number of the figure (in parenthesis), as on p. 87 (see text reference to Fig. 7). It is also unnecessary in the heading to a table or the legend to a figure, to write: Table showing ... *or* Figure showing ... These words should be deleted.

Circumlocution – verbosity – gobbledegook – surplusage – this habit of excess in the use of words, which makes communication more difficult than is necessary, is well established in the speech and writing of educated people:

... of all the Studies of men, nothing may be sooner obtain'd than this vicious abundance of *Phrase*, this trick of *Metaphors*, this volubility of *Tongue*, which makes so great a noise in the World. But I spend words in vain; for the evil is now so inveterate, that it is hard to know whom to *blame*, or *where* to begin to *reform*. We all value one another so much, upon this beautiful deceit; and labour for so long after it, in the years of our education: that we cannot but ever after think kinder of it, than it deserves.
The History of the Royal Society,   Thomas Sprat (1667)

Not all scientists use more words than are needed; nor are scientists

Table 12. Circumlocution: the use of many words where few would do better

Circumlocution	Better English
in virtually all sectors of the environment	almost everywhere
maintain a high degree of activity	move about a great deal
in black and white only	in black and white
if at all possible	if possible
peer groups	equals
I myself would hope	I hope
I would have said	I think
you are in fact quite correct	you are right
mechanisms of a physiological nature	physiological mechanisms
on an experimental basis	by experiment
on a dawn to dusk basis	from dawn to dusk
on a regular basis	regularly
working towards a unanimous situation	trying to agree
by any actual person in particular	by anyone in particular
to show the same high level of application	to keep trying
an oral presentation	a talk
the reading and learning process	reading and learning
outside the kidney itself	outside the kidney
several ... are known to influence	several ... influence
not longer than 20 000 to 25 000 words in length	no more than 25 000 words
measures on purely local terms	local action
a maximum depth of ten metres	ten metres deep
ten metres in length	ten metres long
over a period of the order of a decade	for about ten years
for a further period of fifteen years	for another fifteen years
the roads were limited in mileage	there were few roads
they utilize for sustenance	they eat
during the month of April	in April
at the pre-school level	the under fives
on a theoretical level	in theory
on the educational front	in education
in the classroom situation	in schools
in the school environment	in schools
They are without any sanitary arrangements whatsoever.	There is no sanitation.

Table 12. – cont.

Circumlocution	Better English
...in establishments of a workshop rather than factory character...	...in workshops...
An increased appetite was manifested by all the rats	All the rats ate more
How we speak depends on what speech communities we are actually operating in at the time.	How we speak depends on the people we are with.
It consists essentially of two parts.	It has two parts.
We are in the process of making	We are making
Degree courses are in the process of development.	Degree courses are being planned.
Experiments are in progress to assess the possibility of using	We are trying to use
It was observed in the course of the demonstration that...	We observed...
There is really somewhat of an obligation upon us...	We should...
The committee was obviously cognisant of the problem.	The committee was aware of the problem.
An account of the methods used and the results obtained has been given by ...	Their methods and results are described by
In no case did any of the seedlings develop lesions.	None of the seedlings developed lesions.
Such is by no means the case.	This is not so.
...proved fatal in most cases.	...killed most of them.
Even when the class is engaged in reading and writing activities...	Even when the children are reading and writing...
At the other end of the educational spectrum	In primary schools

more verbose than others. In Psychology and English Studies, many writers consider neither the psychology of their readers nor the use of words (McCartney, 1953).

The Introduction to a Government White Paper on the future of Education in Britain (HMSO, 1972) includes the following circumlocutions:

It lays down the objectives at which the Government are aiming...

Table 13. Circumlocution: some phrases which are commonly used when on
word would do better

Circumlocution	Better English	Circumlocution	Better English
In view of the fact that	because	try out	try
on account of the fact that	as	open up	open
if it is assumed that	if	aimed at	for
in spite of the fact that	although	count up	count
a sufficient number of	enough	check on	check
at this precise moment in time	now	later on	later
at that point in time	then	prior to	before
a greater length of time	longer	seal off	seal
during the time that	while	in between	between
on a regular basis	regularly	inasmuch as	since
it may well be that	perhaps	a number of	several
with the exception of	except	proved to be	were
using a combination of	from	in regard to	about
of a reversible nature	reversible	in all cases	always
which goes under the name of	called	in order that	to
with the result that	so	in most cases	usually
in all other cases	otherwise	at a later date	later
are found to be in agreement with	agree	a proportion of	some
carry out experiments	experiment	a great deal of	much
conduct an investigation into	investigate	at an early date	soon
bring to a conclusion	finish	in the nature of	like
arrive at a decision	decide	not infrequently	often
make an adjustment to	adjust	in the event that	if
make an examination of	examine	to say nothing of	and
undertake a study of	study	has an ability to	can
take into consideration	consider	a small number of	few
afford an opportunity to	allow	a large number of	many
in conjunction with	with	by the same token	similarly
after this has been done	then	for the purpose of	for
on two separate occasions	twice	in the vicinity of	near
the question as to whether	whether	in connection with	about
it is apparent therefore that	hence	until such time as	until
in view of the foregoing circumstances	therefore	spell out in depth	explain
give positive encouragement to	encourage	in this day and age	now
have been shown to be	are	at the present time	now

(The Government's purpose is defined...)

In the 1960s the main determinant of rising educational expenditure was the increasing number of young people using the education system.

(The main reason for the higher cost of education in the 1960s was that more young people were being educated.)

This will call for a sustained co-ordinated effort over a substantial period.

(A sustained co-ordinated effort will be needed.)

Reasons for verbosity

Tautology, circumlocution, ambiguity and verbosity arise from ignorance of the exact meaning of words. Also, people may use too many words (or too few) if they have not considered the difference between speech and writing.

Sometimes in conversation we use more words than are needed in writing. We use words to separate important ideas; we repeat things for emphasis; and we correct ourselves in an attempt to achieve greater precision. These things give the listener time to think.

On the other hand, in conversation, we take short cuts, leaving out words, and so use fewer words than would be needed in writing. This is possible because, as we talk, we see that the listener has grasped our meaning (Fig. 6).

The writer must allow for the lack of direct contact with the reader; and must use as many words as are needed to convey the intended meaning. Emphasis is usually made without repetition, and necessary pauses come from punctuation marks and paragraph breaks.

The writer ... suggests by turns of expression the emphases and gestures of ordinary talk; uses vocabulary that is at once intelligible, interesting and evocative; and so varies his constructions that he avoids the effect of monotony. He gives coherence to speech, at the same time retaining certain of its characteristics.

His immediate appeal is through the eye of the reader, but he does not forget the reader's ear (Vallins, 1960).

Use words with which you are familiar and which match your style to the occasion and to the needs of your readers. Write as you would speak but recognize that good spoken English is not the same as good written English. If a good talk is recorded and then typed, the reader may find that it is not good prose.

In the silences which punctuate conversation
meaning may be conveyed without words.

Fig. 6. *To convey the same information, more words may be needed in writing than in speech.*

Verbosity may be due to confusion of thought, from a failure to take writing seriously, or from laziness in sentence construction and revision. All these things are likely when a report is dictated unless it is revised in typescript. Few people are able to dictate a report, so that it reads well and conveys the intended meaning, unless they spend time in converting the typescript into good prose. If they are prepared to take the trouble, most people should be able to write better than they can talk because in writing there is time for thought and therefore an opportunity for revision. Responsibility for the revision of a typescript cannot be delegated. The work must be checked and changed until the author is satisfied with its content and style.

There are other reasons why people fill their writing with empty words. Some writers seem to think that restatement in longer words is explanation. Others are trying to make a little knowledge go a long way, or they may even be trying to obscure meaning because they have nothing to say, or because they do not wish to commit themselves:

> ... only the wealthy, the capable, or the pretty can afford the luxury of saying right out just what they think, and blow the consequences. *Lieutenant Bones*, Edgar Wallace (1918)

Wordiness may also result from affectation; from the studied avoidance of simplicity (McCartney, 1953) in the belief that Latin phrases, long words and elaborate sentences, appear learned.

Foreign words and expressions ... are used to give an air of culture and elegance. The ends of sentences are saved from anti-climax by such resounding commonplaces as *greatly to be desired, cannot be left out of account, a development to be expected in the near future, deserving of serious consideration,* and *brought to a satis-factory conclusion.* Words like phenomenon, element, individual (as a noun), objective, categorical, effective, virtual, basic, primary, promote, constitute, exhibit, exploit, utilize, eliminate, liquidate, are used to dress up simple statement and give an air of scientific impartiality to biased judgements (Orwell, 1950).

Orwell recommends those who wish to use language as an instrument for expressing and not for concealing thought, to:

1 Be positive. Especially, avoid double negatives such as *not unlikely* (for possible) and *not unjustifiable.*

2 Never use a metaphor, simile or other figure of speech which you are used to seeing in print.
3 Never use a long word where a short one will do.
4 If it is possible to cut a word out, always cut it out.
5 Never use the passive where you can use the active.
6 Never use a foreign phrase, a scientific word or a jargon word if you can think of an everyday English equivalent.

In a technical paper it is not necessary to express complex ideas in language a layman could understand; nor is it necessary to make simple ideas seem complex. Simplicity is the outward sign of clarity of thought. Wordiness is therefore a reflection on a writer's thinking; and a means by which he conceals his meaning – perhaps even from themselves. Reasons for sparing no pains in scholarly writing include fairness to the material, to one's self-respect, and to the reader (McCartney, 1953).

If men would only say what they have to say in plain terms, how much more eloquent they would be.
                    On Style,   Samuel Coleridge (1772–1834)

Any one who wishes to become a good writer should endeavor, before he allows himself to be tempted by the more showy qualities, to be direct, simple, brief, vigorous and lucid.
  The King's English,   H. W. Fowler and F. G. Fowler (1931)

We shall be effective . . . as writers if we can say clearly, simply, and attractively just what we want to say and nothing more. If we really have something worth saying, then we are bound by the nature and necessities of our language to say it as simply as we can.   *Our Language*,   Simeon Potter (1966)

Scholarly writing is so full of verbosity that those who avoid it should be appreciated all the more.

Some scientists believe that objectivity is achieved by writing in the passive voice, or they wish to avoid the undue repetition of personal pronouns. This reluctance to use the first person increases the number of words required and can make writing less rather than more objective. *We found* or *I found* communicate something of the excitement of discovery and make clear who was involved. However, never say *we found* when you mean *I found*. The use of the word *we* (for I) should be reserved for monarchs, editors, and pregnant women. The first person is to be preferred to such expressions as *it was found that*, which may leave the reader wondering who made the discovery. Similarly, it is not always clear who is meant by *the author* or *the writer*.

Almack (1930) wrote: 'Only in the preface is the first person permitted; the remainder of the thesis should in common decency be written in the third person'. McCartney (1953), however, refers to the unwarranted prejudice of authors against the use of personal pronouns. Thornton and Baron (1938) recommend the use of the first personal pronoun for statements of a view that is not generally held, and which is likely to meet with opposition. Kapp (1973) goes further and uses the first person freely: 'I must confess, on re-reading what I have myself written I have frequently caught myself committing the same sin of omission.'

Kapp is referring, in this extract, to the need for an author to include comment words (such as *even*, *dangerously*, *as expected*, and *unexpected*) and connecting words (such as *however*, *hence*, *moreover*, *nevertheless*, and *on the contrary*) to direct the reader's attention. Kapp gives examples of overcrowded writing in which important thoughts are so closely packed that the reader has no time to grasp the full meaning of one before the next is upon him. In practising an economy of words the scientist must not make the mistake of using too few words, so that the connection between ideas is not clear. Reminders should be provided when these are needed, to ensure that the reader always knows what he is supposed to be thinking and why this is relevant. The subject should not be drowned in a sea of words; nor starved of the words needed to give it strength. The rule must be to use the number of words needed to convey a thought precisely (without ambiguity). Brevity must not be achieved at the expense of clarity, accuracy, interest and coherence.

Writing a summary

When you read reports or articles you must recognize the points which are important for your work, or for the interests of your employer. These points may be written on an index card or incorporated in a memorandum.

The following exercises may be completed by anyone working alone; or they may be used by a teacher of scientific writing – first as set work and later as a basis for discussion.

1 Writing a précis is a test of comprehension and an exercise in reduction, in which the essential meaning of a composition is retained – but not the details. The order of presentation should not be changed, unless the order is faulty, and if your précis is carefully made it should require only slight recasting into your own words. Remember that you are not presenting yourself; you are representing the author in brief (Partridge, 1965). It is easier to condense other people's writing than your own but précis writing will help you to develop a clear, simple and straightforward style.

Write a précis of an article or essay written by a colleague or fellow student, or written by yourself some time ago. Include all the essential information and ideas but make the composition as short as you can.

2 In scientific writing, a summary is much shorter than a précis. Select an article that interests you from a recent issue of a journal which does not print an author's summary. Read the article and then read it again as you make notes of the main points. Prepare a summary that is brief (less than 200 words) and yet accurate and informative. Preparing a summary is a good test of your ability to select the most important points, and to report these in a few well chosen words.

3 Select an article from a recent issue of a journal in which authors' summaries are published. The author's summary will serve as a worked example. Cover the author's summary. Read the paper carefully and then prepare your own summary. Do you agree with the author's choice of the most important

points? Has the author used more words than are needed? Have you? The preparation and revision of summaries, like précis writing, will help you to develop a concise and direct style. (See also *The Summary*, p. 138.)

8

Helping the reader

Decide what the reader needs to know
Find out as much as you can about your readers, and then match your vocabulary and style of writing to their needs. Some readers may be experts in the same field, others may be scientists with different interests, and others may know nothing about science, but they all may be concerned in decision-making and in the possible applications of your work. They must understand the parts of your report that are of interest to them.

Editors of scientific journals prefer papers to be written so that scientists working on related subjects can understand them as well as the few specialists working in the same narrow field.

Readers will be most likely to understand and remember your main findings if they can relate anything new to their existing knowledge and interests. Try to anticipate any difficulties so that your writing will be understood at first reading by all those for whom it is intended.

To help your readers, provide an informative title; effective headings and sub-headings; present information in a logical order; include all essential steps in any argument; give sufficient evidence in support of anything new; give examples; and explain why any point is particularly important. No statement should be self evident but you must be as explicit as is necessary, defining every new subject or concept. Do not leave the readers to work out the implications of any statement. Help them to see the logical connection between sentences, paragraphs and sections. Sometimes a word or phrase is enough; sometimes much more explanation is required.

Every word or phrase should be appropriate to its context; and each sentence should convey a whole thought. Fulfil your readers' expectations. For example, follow the words *not only* by *but also*; *whether* by *or*; *on the one hand* by *on the other hand*; and *first(ly)* by *second(ly)*. If you list a number of items, mention all or none of them in the sentences that follow: if only some are mentioned readers may wonder about the others when they should be thinking about your next topic.

Write for easy reading
Your writing should be appropriate to the subject, to the needs of your readers, and to the occasion. Convey your thoughts clearly, accurately and impartially so that your readers take your meaning and always feel at ease.

How to begin
If you know what you wish to communicate but have difficulty in getting started, look at the opening sentences of similar compositions by other people. Begin, for example, with a summary; with recommendations; a statement of the problem; a hypothesis; necessary and interesting background information which leads directly to the problem or hypothesis; an example; a definition; a question; an answer to one of the reader's six questions (who, what, where, when, why or how); a new idea; an idea which has received some support (then explain why it is incorrect); an accepted procedure (then explain the advantages of another procedure). The best starting point, for the subject and for your readers, may become apparent as you prepare your topic outline. It is better to get started, however, than to spend too much time trying to think of an effective beginning. Your introductory paragraph can be revised, if necessary, when your first draft of the whole composition is complete. The only rules about beginning are: come straight to the point; and, if possible, refer to things that you expect your readers to know and build on this foundation.

Control
This depends upon your knowledge of the subject and upon

careful planning – which makes possible the presentation of your knowledge in an appropriate, ordered and interesting way. Good headings and sub-headings, especially in a long composition, are signposts that help the readers along and help them to know where to look for the information they require.

Emphasis

The title, the headings and the sub-headings emphasize the whole and its parts. Emphasis, which is achieved in many ways, is important in all writing, and is present whether or not the writer is in control. But you can use emphasis effectively only if you know how to make important points stand out from the necessary supporting detail.

Beginnings and endings are most important. The first and last paragraphs (the introduction and the conclusion) will be read by most people. The most important words in each paragraph are the first words (so miss out unnecessary introductory phrases, see Table 7) and the last words (so end each paragraph effectively). The most important parts of a sentence are the first and last words.

Items of comparable importance may be emphasized by repetition of an introductory word, by numbering, or by indentation. If a sentence is well balanced, so that it reads well, emphasis will fall naturally on each part.

Leave out anything that is irrelevant, *details* that your readers already know, and any unnecessary background information. Use more forceful language for important points than for the supporting detail; repeat important words; and plan effective illustrations to convey the essential points in your composition.

Mark the points you wish to emphasize in your topic outline, and check your first draft to ensure that they are sufficiently emphasized.

Sentence length

Long involved sentences may indicate that you have not thought sufficiently about what you wish to say. Revise any long sentence if it is difficult to read, or make it into a number of shorter sentences. Match your sentence length to the needs of the reader. The breaks between sentences give time for thought and Flesch (1962) grades writing, according to *average* sentence length, as very easy to read (less than 10 words), difficult (more than 20 words) and very difficult (30 words).

Although short sentences are the easiest to read, a long sentence, if it is properly constructed, may be easier to read than a succession of short ones; and there is no rule that a sentence, when read aloud, should be read in one breath. Good prose is seldom written in short sentences. Sentences vary in length. Short sentences are effective for introducing a new subject, long sentences in developing a point, and short ones in bringing things to a striking conclusion.

'If you really want to know,' said Mr. Shaw with a sly twinkle, 'I think that he who was so willing and able to prove that what was, was not, would be equally able and willing to make a case for thinking that what was not, was, if it suited his purpose.' Ernest was very much taken aback.

The Way of All Flesh, Samuel Butler (1903)

Rhythm

This is proper to verse and is rarely absent from effective prose. Prose has its origin in speech and is often a record of speech. It has a varied rhythm that conveys a certain pleasure to the ear, and yet contrasts with the strict metred rhythm of verse. In prose, the forward march of language, the varied rhythm contributes to the flow of words in a sentence. This, with the flow of logically arranged ideas in successive sentences, helps to make a passage interesting and easy to read. Rhythm may give emphasis, help to present shades of meaning, or ensure that communication is achieved freely and pleasurably. This is not to say that the writer deliberately sets out to introduce rhythm; he relies on the natural tact of the ear (Henn, 1960).

Use punctuation marks to clarify meaning and to contribute to the smooth flow of language. Effective prose usually sounds well, and a good test of your writing is to read it aloud and to

revise any parts which do not sound well. Writers with a feeling
for the sounds of words try not to offend the ear (McCartney,
1953):

1 by unintentional alliteration, as in *with bundle bands bending*,
   and *rather regularly radial*;
2 by the grating repetition of s, as in *such a sense of success*;
3 by adding s to a word that does not require it, such as *toward*
   and *forward* (but the s may be needed to make a sentence
   easier to read);
4 by the repetition of syllables, as in *appropriate approach*,
   continue to contain, and *protection in connection with infection*;
5 by the repetition of sound, as in *found around*, *elaborate
   laboratory facilities*, and *with respect to the effect*;
6 by the repetition of cognate forms in different parts of speech,
   as in *a locality located*, *the following procedure should be
   followed*, and *except for rare exceptions*; and
7 by repeating a word with a change of meaning, as in *a point
   to point out*.

Style

Some scientists may feel that style is not important in scientific
writing. But style is not an extra – added to writing as a final
polish – it is part of effective prose. The word style is, moreover,
difficult to define, for definitions are like sieves through which
'the particular achievement of genius is so apt to slip' (Quiller-
Couch, 1916). However, Swift defined style as *proper words in proper
places*; Newman said that style is a *thinking out into language*; and
Matthew Arnold considered that the secret of style is *to have something to say and to say it as clearly as you can*.

Graves and Hodge (1947) suggest that the style of prose best
suited to the second half of the 20th century, should be:

Cleared of encumbrances for quick reading; that is, without
unnecessary ornament, irrelevancy, illogicality, ambiguity, repetition, circumlocution, obscurity of reference.
Properly laid out; that is, with each sentence a single step and

each paragraph a complete stage in the argument or narrative; with each idea in its right place in the sequence, and none missing; with all connections properly made.

Written in the first place for silent reading, but with consideration for euphony if read aloud.

Consistent in use of language; considerate of the possible limitations of the reader's knowledge; with no indulgence of personal caprice nor any attempt to improve on sincere statement by rhetorical artifice.

The need for careful planning is emphasized in these notes on style, and in George de Buffon's address to the Académie Française in 1703:

> This plan is not indeed the style, but it is the foundation; it supports the style, directs it, governs its movement, and subjects it to law. Without a plan the best writer will lose his way. His pen will run on unguided and by hazard will make uncertain strokes and incorrect figures. Style is but the order and the movement that one gives to one's thoughts.

In writing about science, a good style depends upon your intelligence, imagination and good taste; upon sincerity, modesty, careful planning, and attention to the requirements of scientific writing. You are familiar with these things as part of the scientific method. In effective prose the excitement of discovery may also be communicated. Rhythm, while not essential, will make for easier reading, and badly constructed sentences may irritate readers and make them less receptive to your message.

Capture and hold the reader's interest
In writing about science in a book, in an article for a journal, in a project report, or in describing an experiment, your interest in your work should be conveyed to your readers.

A novelist, whose business is words, must quickly capture and hold the reader's interest. He takes great care over the choice and use of words. Consider, for example, the first paragraph of a successful novel:

He rode into our valley in the summer of '89. I was a kid then, barely topping the backboard of father's old chuckwagon. I was on the upper rail of our small corral, soaking in the late afternoon sun, when I saw him far down the road where it swung into the valley from the open plain beyond.

Shane,   Jack Schaeffer (1954)

The first two words capture the reader's attention. The first sentence (in ten short words) tells what the story is about; it begins to answer the reader's questions – who, where, and when? The first paragraph tells that the story will be told as it affected the life of a boy. No word is superfluous. Each one plays a part in setting the scene.

Scientific writing is more direct, but your intention is the same. You start with the advantage that your readers are interested but you must maintain this interest. Present information at a proper pace. If readers understand they will want to move quickly to the point. However, they must understand every word, every statement, and every step in an argument; for if they must consult a dictionary or read a sentence again, to confirm that they have taken the right meaning, their attention may be lost.

Readers are directed away from your explanation or argument by anything that is not relevant, by unnecessary detail, by the explanation of the obvious, or by needless repetition. They will lose interest if statements are not supported by sufficient evidence.

Science depends upon evidence and you must not attempt to gain acceptance of your views by their reiteration. Cross references can be used to avoid repetition and to provide necessary reminders. When anything is repeated, for emphasis or to help to clarify a difficult point, use a phrase such as *that is to say* or *in other words*. Otherwise, after studying both sentences to make sure that their meaning is the same, readers may still wonder if they have failed to appreciate some difference.

Approach people through *their* interests rather than your own. They will be most interested in themselves, in other people, and in things as they affect people. Scientists are most interested in their own speciality, and in developments in other fields which may have a bearing upon their own work. The man-in-the-street is likely to be most interested in the application of science to human welfare, in the impact of science on society, and in pure research when this is concerned with man's origins.

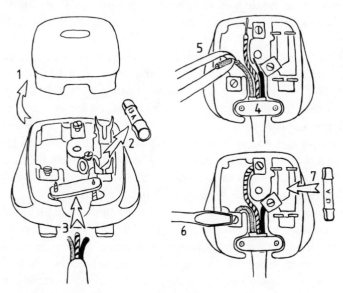

Fig. 7. *Strip cartoon used to convey information, either instead of writing or to support the words.*
Drawing by Dave Douglas.

In a specialized scientific journal the reader is interested in methods and in detailed results. The style of writing is usually direct and the link between paragraphs is achieved by their logical connection. In a scientific magazine, which has a wider readership, more explanation and interpretation are needed. A newspaper must interest people who would not look at a scientific magazine – by reference to familiar things and by examples, anecdotes, analogies and attractive illustrations. A strip cartoon (Fig. 7, for example) may inform people who would not read an article on the same subject.

Use good English

Readers should not have to waste time trying to understand things which are badly expressed (Table 14).

If your work is to be published, remember that readers who do not speak English as their mother tongue are most likely to understand grammatically correct English supported by effective illustrations.

To write well most people need to be alone, free from disturbance, and with time for thought. Poor writing may result from distraction, from not knowing what to say, from not considering how to present the information, from insufficient care in the choice and use of words, or from leaving too little time for writing, correction and revision. Poor writing is also to be expected from a writer who has nothing to say, or who does not wish to express an opinion, and is so inconsiderate as to try to put up a smoke-screen of words which sound well but say nothing.

During long periods of research a scientist may have no practice in organizing a larger composition (but see p. 146) and then may be unwilling to devote enough time to writing. 'Compelled at last to put pen to paper, he discovers (or, more accurately, his reader does) that he has lost or never acquired the necessary techniques, which means that, before the manuscript can be printed somebody must patch it' (Gill, 1954).

Some scientists do not write as well as they should because they do not think of writing as part of science. If they then present their results and conclusions in language which is difficult to read, they may find that their work is ignored. If they fail to make their meaning clear, misunderstandings may cause other people wasted effort and may lead to costly mistakes.

Looking critically at other people's writing (see p. 34 and Tables 14 and 15) will help you to improve your own, but do not be afraid to put your pen to paper for fear of making mistakes. You should, rather, take comfort from the mistakes of even experienced writers. It is difficult to write well consistently, and Vallins (1964) gives extracts from the work of well known authors to show how many traps there are in English. Recognizing them ...

> gives us a deeper knowledge of and a keener sensitiveness to language; so that in our own more humble, more ordinary writing, we may steer clear of the more elementary errors because we have learnt to recognise subtler ones in the writing of other people.

There is no need to be tied to the conventions of purists in the use of English. Our language is not fixed for all time. Changes are made by the people, who must use the language of their day (standard English or standard American). While there is no correct English or plain English (or even standard English in the sense of a language that never changes), educated people should try to speak and write good English (Vallins, 1964).

Young writers should be encouraged by the thought that English is bad only if it does not express clearly and accurately, in words appropriate to the context, the thought intended (Vallins, 1964). Prose should be judged by the ease with which it conveys its full intended sense to the readers to whom it is addressed, rather than by its correctness by the laws of formal English grammar (Graves and Hodge, 1947). This is not to say that grammar (the art of speaking, reading and writing correctly) is unimportant. Mistakes in grammar make writing inaccurate, imprecise and ambiguous.

Obstacles to effective communication, include:
1 lack of practice on the part of the writer;
2 the writer's unwillingness to spend enough time in thinking, planning, writing and revising;
3 failure to establish contact with the reader at the start;
4 lack of attention on the part of the reader, especially when the writing deviates from his own speciality;
5 the reader's preconceived ideas, and his refusal to accept new ideas or to consider evidence that conflicts with his existing beliefs.

Table 14. Extracts from books and from published papers

Extract	Suggested improvement
More than mere numbers are required.	More is required than numbers.
Mercuric chloride gave the best control of the seven materials used.	Of the seven materials used, mercuric chloride gave the best control.
Attempts to isolate ... from ... gave negative results.	We were unable to isolate ... from ...
Discussions with colleagues in other laboratories elucidated that this problem is not uncommon.	We know, from talking to people in other laboratories that they have this problem.
The discovery of isotopes which spontaneously emit radiation was first made more than 80 years ago and since that time ...	Isotopes which spontaneously emit radiation were discovered over 80 years ago, and since then ...
Southern industry was much more limited in scope and structure. Much of it was of artisan character whose inefficient and relatively high cost of production was only possible under the shelter of a high tariff wall and a paucity of communications that gave almost a monopoly in local markets.	Many firms in the south were inefficient. Their production costs were high but, because of poor communications within the country and high tariffs at the ports, they had few competitors in local markets.
Human beings hope to survive, and be comfortable, on this planet for a length of time that far exceeds the duration of all fuel reserves. Physicists therefore are now applying themselves to the question of whether science can possibly harness alternative sources of energy.	We hope to survive in comfort on this planet for longer than fuel reserves will last. This is why physicists are asking: 'Can we harness other sources of energy?'

Table 14. – cont.

Extract	Suggested improvement
The scope and magnitude of tasks that need to be performed in hydrospace have expanded rapidly in recent years principally due to the growth in extraction of hydrocarbons from offshore fields. This has resulted in nearly 6400 km of pipeline and 100 permanent production platforms in the North Sea alone.	More and bigger tasks must be performed in the sea, especially in the offshore oil and gas fields. In the North Sea, for example, a hundred production platforms and nearly 6400 km of pipeline have been constructed.
Assuming that one is dealing with a sufficient number to constitute a reasonable group or series, a 50 per cent proportion of failure is as much a failure of the instructor as of the students, because it indicates not only inability to stimulate a sufficient amount of mental activity but also a faulty sense of proportion.	If this is a representative sample, a 50 per cent failure rate indicates that the instructor was unable to make the students work and that ...
It is a sort of aesthetic problem. We could, of course, proceed on the assumption that ...	This is an aesthetic problem. We could assume that ...
They all have a very similar chemical formula and are similar in their physiological properties.	They all have similar chemical formulae and similar physiological effects.
The discovery of hormones also contributed to increased food production.	The use of hormones has contributed to increased food production.
33 kg per hectare of nitrogen	33 kg nitrogen per hectare
increasing the average size of individuals	increasing the body size of individuals

Table 14. – cont.

Extract	Suggested improvement
A group of men rendered *signal* service in *diffusion* of knowledge of physics *were* the scientific instrument makers. *It goes without saying that* without them research in science would have been slowed down: scientific instruments are *indispensable allies* in scientific discovery. It was only after the *discovery* of the microscope that such sciences as histology and microbiology could begin. It was only after apparatus for the *creation* of high vacua had been perfected that the study of electrical discharges in rarefied gases became possible. It was only with ... invention of the cloud chamber, which made visible the track of a charged particle, that subatomic physics could make another *leap* forward. *Mention must be made* of men such as Faraday, Rutherford and Bragg, who not only made momentous discoveries but built their own apparatus to make their discoveries.	Without scientific instruments many discoveries would not have been made. Microbiology and histology were made possible by the invention of the microscope; the study of electrical discharges in rarefied gases by the construction of a high vacuum apparatus; and the tracking of charged particles by the invention of the cloud chamber. Faraday, Rutherford, Bragg, and others, made discoveries with apparatus that they themselves had built.

Note that the words printed in italics in the extract are either incorrect or are superfluous introductory phrases.

Table 15. Advice on the use of language

Extract	Suggested improvement
The purpose of these general notes is first to assist authors in the writing of scientific papers in an acceptable style and secondly to suggest the introduction of greater uniformity in the general approach to the preparation of scientific papers for publication.	These notes are to help authors to write scientific papers in an acceptable style, and to encourage greater uniformity in presentation.
It is, nevertheless, clear that specialized subjects may call for particular methods of presentation, and these notes should be read in conjunction with any instructions issued by the journal in which the author hopes to publish. When the notes conflict with such instructions, the latter should be followed.	But, if these notes conflict with the editorial policy of the journal in which you hope to publish, the rules of the journal must be followed.
... we must convince the teacher of history or of science, for example, that he has to understand the process by which his pupils take possession of the historical or scientific information that is offered them; and that such an understanding involves his paying particular attention to the part language plays in learning.	... teachers of history or of science, for example, must understand how their pupils learn history or science, and they must pay particular ...
If a man who sets out to write does not show some respect for his medium, he immediately condemns himself in the eyes of the people who have to read what he has written.	The writer who shows no respect for language condemns himself.
So listeners tend to send me questions about words, ...	Some listeners send me questions about words, ...

Rules for efficient communication

1 Always decide what you wish to say, why you wish to say it, and whom you hope to interest, before you start to write.

2 Write of things you know, if you have something interesting to say.

3 Plan your work so that information and ideas can be presented in a logical and effective order, and so that the whole composition has the qualities of balance and unity.

4 Write for easy reading. Begin well. Keep to the point. Be clear, direct and forceful. Maintain the momentum of your writing to the end. End effectively.

5 Revise your work (see p. 43 and p. 147).

In starting to play any game it is always a good idea to watch an expert. In learning to write effectively, study the technique of the successful writer. Just as the way we speak is influenced by the speech we hear, so in learning to write we are influenced by the things we read.

The King James I Bible had a profound influence on speech and writing when it was the staple literature of English speaking people, just as the daily newspapers do today. Most people read newspapers, and some read only newspapers. Journalists and broadcasters use the English of today and exert an untold influence on the development of the language. The young writer should study the technique of journalists who write well (in leading articles, for example) to see how to capture the reader's attention, how to match the writing to the needs of the reader, and how to write clear, concise, vigorous and vivid prose.

A simple, straightforward style is required in scientific writing and young people who are still developing a style of their own will find such direct prose in the works of, for example, John Buchan, Winston Churchill, Daniel Defoe, Aldous Huxley, Robert Louis Stephenson, and Jonathan Swift. It is a mistake, however, to attempt to copy someone else's style. There is no one correct way to write, since the way each person puts words together to convey meaning reflects his personality and his feeling for words.

Leonard was trying to form his style on Ruskin: he understood him to be the greatest master of English Prose. He read forward steadily, occasionally making a few notes.

'Let us consider a little each of these characters in succession, and first (for of the shafts enough has been said already), what is very peculiar to this church – its luminousness.'

Was there anything to be learnt from this fine sentence? Could he adapt it to the needs of daily life? Could he introduce it, with modifications, when he next wrote a letter to his brother, the lay reader? For example:

'Let us consider a little each of these characters in succession, and first (for of the absence of ventilation enough has been said already), what is very peculiar to this flat – its obscurity.'

Something told him that the modifications would not do; and that something, had he known it, was the spirit of English Prose. 'My flat is dark as well as stuffy.' Those were the words for him.   *Howards End*,   E. M. Forster (1910)

9

Numbers contribute to precision

A politician may say that he firmly believes that a fund will be established 'of *substantial* size and *adequate* coverage over a *considerable* period'. He uses vague words to express his hopes when he is unable to be precise. In science precision is needed and yet 'nearly all scientists, at the point where they turn from mathematical or chemical language to English, seem to feel relieved of any further obligation to precise terminology' (Graves and Hodge, 1947).

Consider the meaning that you wish to convey before you use the word *very* as an adverb (*very quickly*) or as an adjective (*very large*), and before you use other adjectives (such as *small, light, appreciable, large* and *heavy*) or modifying and intensifying words (such as *actually, comparatively, exceptionally, extremely, fairly, quite, rather, really, relatively* and *unduly*). Vague statements will annoy the reader:

> Whenever anyone says I can do something *soon* I'll say to them, yes, I know all about that ..., but when, when, when?
>
> *Key to the Door*,   Alan Sillitoe (1969)

Numbers convey more information and contribute to precision.

The use of numbers
In writing, cardinal numbers (twenty-one to ninety-nine) and ordinal numbers (e.g. twenty-first; one-hundred-and-first) should be hyphenated.

Use words, not figures, at the beginning of a sentence, and for numbers one to nine. Otherwise, figures should normally be used for numbers greater than nine, and they must be used before a symbol.

Two numbers should not be given together, either as numerals or in words, or ambiguity may result. Write two 50 W lamps, not 2 50 W nor two fifty watt.

Where necessary, percentages should be defined (e.g. in describing solutions percentage by mass must be distinguished from percentage by volume).

Decimals are indicated by a full stop on the line in English language journals and by a comma in some languages, not by a point raised above the line (3.2 or 3,2 but not 3·2). The comma should not therefore be used to break numbers above 999 into groups of three digits. With more than four digits, spaces should be left: 9999 or 10 999 or 999 999 not 9,999 or 10,999 or 999,999. Where possible prefixes and symbols should be used to indicate decimal multiples and submultiples and prefixes involving powers of three are to be preferred (Table 16). Because of differences between European and U.S.A. usage, the words billion, trillion and quadrillion should not be used.

The use of decimals should contribute to accuracy in measurement. The numbers 5 and 5.0 and 5.00 indicate different degrees of precision. For values less than one a zero should be placed before the decimal point (0.25 not .25). Modifying words such as about, more or less should not be used with decimals. Remember that the results of a calculation should not be expressed in more places of decimals than are present in the least accurately known component of the calculation; otherwise your results will appear to be more precise than is possible with the method of measurement used in obtaining the data. Remember, also, that accuracy in calculation can do nothing to compensate for lack of care in the collection and recording of data.

Original data are not usually presented in the body of a report, but are summarized in graphs, diagrams or tables, or described by statistics in the text. The methods used in statistical analyses should be indicated; the symbols should be those used in a recent and authoritative book on statistics, and their meaning should be explained.

Differences which are not statistically significant should not be described as insignificant, and scientists interpreting the results of statistical analyses should remember that when something is improbable this does not mean that it will never happen. When things occur in sequence the first is not necessarily the cause of the second; and when two things are shown to be correlated this does not mean that one is necessarily the cause of the other.

For further guidance on the use of numbers, see *Quantities, Units and Symbols* (Royal Society, London); *Changing to the Metric System* (HMSO, London) for conversion factors, symbols and definitions; and the list of Standards (p. 123).

Table 16. Multiples and submultiples
Prefixes and symbols used with SI units to indicate decimal multiples and submultiples. Prefixes involving powers of three to be preferred.

Multiples			*Submultiples*		
Factor	*Prefix*	*Symbol*	*Factor*	*Prefix*	*Symbol*
10^{18}	exa	E	10^{-1}	deci	d
10^{15}	peta	P	10^{-2}	centi	c
10^{12}	tera	T	10^{-3}	milli	m
10^{9}	giga	G	10^{-6}	micro	μ
10^{6}	mega	M	10^{-9}	nano	n
10^{3}	kilo	k	10^{-12}	pico	p
10^{2}	hecto	h	10^{-15}	femto	f
10	deca	da	10^{-18}	atto	a

SI units
See Table 17 and footnote. The magnitude of any physical quantity must always be stated as the product of a pure number and an SI unit (Physical quantity = number × unit). The following rules should be followed in the use of symbols with SI units.

1 Leave a space between the numerical value and the unit (50 W).

2 Do not use a full stop after a symbol unless it comes at the end of a sentence.

3 Never add s to a symbol: m = metre or metres.

4 Do not leave a space between a prefix and a symbol: millisecond = ms.

5 Leave a space between the symbols when two or more symbols are combined to indicate a derived unit: metres per second = m s^{-1} (or m/s). Acceleration is indicated as m/s^2 (not as m/s/s).

6 Do not leave a space between the degree sign and the letter C but leave a space between the degree sign and the preceding numeral: 20 °C not 20°C nor 20° C.

7 Symbols for physical quantities are printed in italics. Symbols for units are printed in Roman type. If, on a graph, potential difference (V) measured in volts (V) is to be plotted against current (I) in milliamps (mA), the axes should be labelled:

either V/V   and I/mA
or V in volts and I in milliamps

8 Symbols for vector quantities are printed in bold face italic type (e.g. *F* for force). Symbols for tensors of the second order should be printed in bold face sans serif italic type (e.g. **S**).

The use of tables

Tables, like illustrations, should be as clear and simple as possible. They should be summaries of your results, as an aid to interpretation, and might facilitate, for example, comparison of the arithmetic means of different samples. Long tables of data, if they are needed in a report, should normally be in an appendix.

Tables are useful because they include summaries of relevant statistics but do not interrupt the flow of words in the text. The information in a table should not be repeated in the text, or in an illustration. Nor should a table include columns of numbers if

Table 17. International System of Units (SI units)

Quantity	Unit	Symbol
length	millimetre (0.001 m)	mm
	centimetre (0.01 m)	cm
	metre	m
	kilometre (1000 m)	km
area	square centimetre	cm^2
	square metre	m^2
	hectare	ha
volume	cubic centimetre	cm^3
	cubic metre	m^3
capacity	millilitre (0.001 l)	ml
	litre	l
mass	gramme (0.001 kg)	g
	kilogramme	kg
	tonne (1000 kg)	t
density	kilogramme per cubic metre	kg/m^3
time	second	s
	minute (60 s)	min
	hour (3600 s)	h
	day (86 400 s)	d
speed, velocity	metre per second	m/s
	kilometre per second	km/s
plane angle	radian	rad
solid angle	steradian	sr
frequency	hertz	Hz
force	newton	N
pressure	pascal	Pa
energy, work, quantity of heat	joule	J
electric current	ampere	A
power, energy flux	watt	W
	kilowatt	kW
electric charge	coulomb	C
electric potential	volt	V
electric resistance	ohm	Ω
electric conductance	siemens	S
electric capacitance	farad	F
magnetic flux	weber	Wb
magnetic flux density	tesla	T
inductance	henry	H

Table 17. – cont.

Quantity	Unit	Symbol
luminous flux	lumen	lm
illuminance	lux	lx
luminous intensity	candela	cd
luminance	candela per square metre	cd/m^2
thermodynamic temp. (T)	kelvin	K
Temperature (t)	degree Celcius	°C
amount of substance	mole	mol
concentration	mole per cubic metre	mol/m^3

Notes: In the International System of Units, the metre, kilogramme, second, ampere, kelvin, candela and mole, are *base units*. Other units, such as the centimetre and kilometre, are *derived units*, being submultiples or multiples of base units. The radian and steradian are supplementary units (not classified as either base units or derived units). The litre, tonne, minute, hour, day, and the degree Celcius (but not the micron) are recognized units outside the International System. The hectare is accepted temporarily in view of existing practice. In Britain the degree Celcius is called the degree Centigrade. For further information on SI units, including units not shown in this table, see list of Standards on p. 123.

these can be calculated easily from numbers in other columns.

The heading should indicate clearly and concisely what each table is about. It should be possible to understand the table without reference to the text, but there should be further explanation in the text and each table should be in the most appropriate place.

The stub of a table (the first column on the left) identifies the horizontal lines. Sub-headings may also be required if a table is in several parts. Each column should have a concise heading in which units are stated for every quantity shown; and, as appropriate, the prefixes listed in Table 16, p. 98. If there is no entry in any part of a table, this should be shown by three dots ... and a footnote to indicate that no information is available. Use an o for a zero reading only. Decimal points in a column must be in a vertical line.

In a table, comparison should be possible both vertically and horizontally; and where a total is given in the bottom right hand corner, the vertical and horizontal totals must agree. If a table is to be published, consult a recent issue of the journal or the house rules for the usual practice in the use of horizontal and vertical lines. Some journals require Roman numerals for the numbers of tables and Arabic numerals for illustrations. Others require Arabic numerals for both. The tables should be numbered consecutively, separately from the illustrations.

Decide the size and shape of each table in relation to the size and shape of the page or column in which it is to fit. If possible the table should fit upright on the page (or column). Each table should be on a separate sheet, with any necessary footnotes below the table (but with no text matter on the same sheet). Each footnote should be preceded by a letter or symbol (not by a number) which must also be shown in the table to identify the entry to which the footnote refers.

The use of graphs and diagrams

Graphs

These are used for presenting numerical data or results derived by the analysis of data, in such a way as to facilitate comparison. The points on a graph may therefore be either single measurements (as in a scatter diagram) or average values. In the latter, either the standard error ($\pm S\bar{x}$) or the 5 per cent fiducial limits of error in the mean ($\pm 1.96\ S\bar{x}$) may be shown by a vertical line through the mean.

There are conventions that must be followed in the preparation of graphs (see also p. 116). In a two-dimensional graph the horizontal axis (the x axis or abscissa) and the vertical axis (the y axis or ordinate) cross at right angles. On the x axis, positive values are shown on the right of the y axis and negative values on the left. On the y axis, positive values are shown above the x axis and negative values below (Fig. 8). Each point on the graph is described by two numbers, its coordinates, which give its position with respect first to the x axis and then to the y axis.

A graph shows how one thing varies relative to changes in another. One thing may be controlled and varied (temperature for example) while some effect of this change is measured; alternatively an effect of changes in some uncontrollable factor such as time may be measured. Temperature and time, in these examples, are independent variables which are always plotted in relation to the horizontal axis. The effect these changes have on something else (the dependent variable) is plotted in relation to the vertical axis.

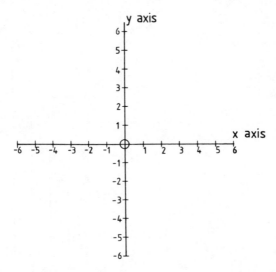

Fig. 8. *The parts of a graph.*
The thing that you can control must be plotted in relation to the *x* axis.

Information should not be repeated, for example in a table and in a graph, in the same report. A graph may occupy more space than a table but, for some purposes, a graph is better because of its immediate visual impact. A graph, like a table, should have a heading, or it should have a legend (see p. 117 *Writing the legend*).

The artful advertizer may misuse graphs in an attempt to mislead people. Scientists should take care not to do this. To help the reader, when two or more graphs are to be compared they should be drawn to the same scale and if possible they should be side by side. The scales used for the axes of a graph should normally start from zero; they should be chosen carefully and marked clearly. Units of measurement must be stated (see rules for use of symbols with SI units, p. 99) and any break in the scale should be indicated. All numbers on the scales should be upright but the labelling of the scales should be parallel to the axes.

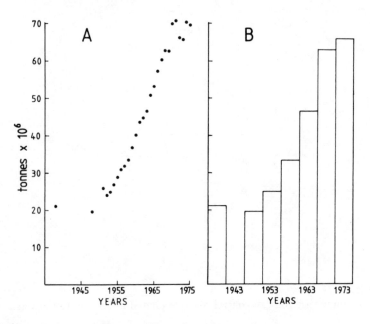

Fig. 9. *Fishery statistics: world catch 1938 to 1975.*
Presentation of data in (A) a graph (all available figures); and
(B) a histogram (catch for every fifth year). All estimates from
Yearbook of Fishery Statistics, FAO, Rome.

A false impression may be given if successive points on a graph are connected by lines (as in Fig. 16, p. 126). It may be, for example, that any trend could be represented better by a best-fitting straight line or curve, or it may be better to present the information as a histogram (Fig. 9B), or to leave the points on the

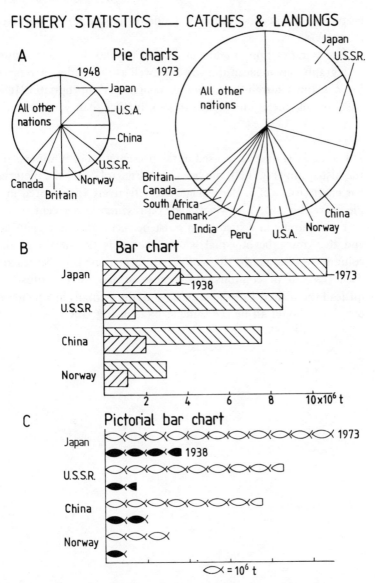

Fig. 10. (A) *Pie charts;* (B) *bar chart;* and (C) *pictorial bar chart.*
All estimates from *Yearbook of Fishery Statistics*, FAO, Rome,
except that no estimate is given for China in 1938 and 1948.

graph with no persuasive line to guide (or misguide) the observer
(Fig. 9A).

The extrapolation of a line or a curve beyond the points shown
on a graph may mislead the writer as well as the reader. A remark
by Winston Churchill, in another context, is appropriate. 'It is
wise to look ahead but foolish to look further than you can see.'

Diagrams
Other kinds of diagrams, used in the presentation of numbers, are
the histogram (Fig. 9B), the bar chart (Fig. 10B) and the pictorial
bar chart (Fig. 10C). Percentages are effectively represented in a
pie chart (Fig. 10A) in which 360° represent 100 per cent.

The columns in a histogram must be rectangles, not pictures,
and they must be of equal width since it is the height of the
column (not its width) that conveys information. If drawings are
used (instead of columns) differences in area may confuse or
mislead the reader. For the same reason, the symbols in a pictorial
bar chart should all be the same size (as in Fig. 10C).

10

Illustrations contribute to clarity

Pictures attract attention and help you to present information quickly, concisely, clearly and accurately. They stimulate and interest the reader, and they should help and inform.

> And ye who wish to represent by words the form of man and all the aspects of his membranification, get away from the idea. For the more minutely you describe, the more you will confuse the mind of the reader and the more you will prevent him from a knowledge of the thing described. And so it is necessary to draw and describe.
>
> *Notebooks*, Leonardo da Vinci (1452–1519)

Illustrations should complement your writing. Do not consider them as ornament, or as additions to a work that is otherwise complete. Your composition should be planned so that information is presented in the most appropriate way: in words and pictures. You should think of preparing rather than of writing a paper.

The use of illustrations

Each illustration should be numbered, so that it can be referred to in any part of the composition, but it should be in the most appropriate place. The reader should be able to look from the text to the figure: as far as possible your results should be presented and allowed to speak for themselves.

A sequence of illustrations, like stills in a film strip or in a strip cartoon, may give a good indication of what the text is about; and anyone looking at a well illustrated article written in another language may look first at the illustrations which provide an

PARTS OF A FLOWER

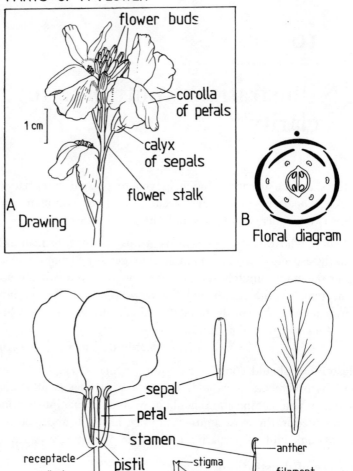

A Drawing

flower buds

corolla
of petals

calyx
of sepals

flower stalk

1 cm

B Floral diagram

sepal

petal

stamen

receptacle

pedicel

pistil

stigma
style
ovary

anther

filament

1 cm

C Diagram: cut away and explosion technique

Fig. 11. *The drawing (A) and the diagrams (B and C) are half
the dimensions of the original artwork. Lettering with 0.7 mm
and 0.5 mm pens; letters 7 mm high (capitals) and a maximum
height of 7 mm or 5 mm for lower case letters, prior to reduction.
Other lines drawn with a mapping pen.*

international language. If you wish to convey a message so that it can be understood by as many people as possible you will prefer pictures to words.

Illustrations are an aid to precise description. They should provide information needed for an understanding of the work and should, therefore, reduce the number of words required in the text. If its message is simple and clear, a brief reference to the figure may be all that is needed in the text (see p. 69).

Illustrations should make an immediate impact. They should not be included as a form of padding; nor should authors be deterred by production costs from including a useful illustration. For a printed report the cost of block making should be offset by the fewer words needed in the text.

Line drawings are carefully planned to fit the needs of the composition and authors are unlikely to include too many but there may be a temptation to include a photograph because it is a good one. Photographs should not be used as ornament but should, if possible, be taken to support the text.

Many people have an uncritical respect for the printed word and too readily accept what is written as necessarily true. The scientist learns to recognize differences of opinion and to read critically. However, illustrations have an immediate impact. No one reads without first choosing to do so but a glance at a picture may leave a lasting impression. This is why advertizers prefer pictures to words.

The value of illustrations cannot be over-emphasized but, because they are so effective in conveying a message, they must be planned and produced carefully so that the reader is not misled. The same test of clarity and truth must be applied to your illustrations as to your writing.

Consider the reason for each illustration and what information you wish to convey before you decide what kinds of illustrations to use.

Photographs
These serve the double function of depiction and corrobora-

tion. Their use in support of scientific writing is accepted. They enable the reader to see what is described in the text.

However, the reader may be too easily convinced that what he thinks he sees in a photograph is true; that is to say, that his interpretation is correct. A photograph cannot lie but it may mislead. This is especially likely when natural shadows, which give a three-dimensional effect, are destroyed by artificial lighting. Furthermore, when a report is to be published even the best prints lose something in reproduction (and they may also be half the dimensions of the original). As a result, the things you see in your original print may not be seen by the reader – who may not know that they are there. Many authors are disappointed at the loss of detail when the plate is compared with the photograph.

A line drawing may be better than a photograph for illustrating microscopic or very small objects. In the photograph much of the subject may be out of focus but the microscope can be focused in different planes at different stages in the preparation of a drawing to bring out important details.

In selecting photographs, look for scientific interest, sharpness of focus, effective lighting and contrast, and then consider whether or not a good line drawing or a diagram would better serve your purpose.

Drawings

These are not intended as proof but as illustration. The burden of proof rests on the scientist. In a line drawing you can help to avoid confusion by directing emphasis to those things you consider essential to your argument.

Each line in a drawing should be an accurate record of an observation. Because of this, drawing is both an aid to observation and a summary of observations. If the proportions are to be correct the drawing must be to scale, and the scale should be marked on the drawing in metric units. When a number of drawings are part of one figure they should, preferably, be drawn to the same scale, as should the figures for one article.

Inaccuracies in a drawing may result from lack of knowledge of the subject but this is usually corrected by the careful study required for the preparation of the drawing. Drawing is therefore a useful part of a scientist's training in the art of observation.

Inaccuracies may result when a subject is so well known to the artist that he represents what he believes to be true rather than what he can see. A student's drawings, for example, may include details, from a text book, which he has not seen in his practical work. The experienced scientist must also be careful to draw what he sees rather than what he believes by experience to be there.

Drawings and photographs have several things in common. They represent three-dimensional objects in two dimensions. They represent things as they are seen at one instant from one place. The photograph is fixed whereas the things photographed may change, and even when the photograph is taken the subject may look different from another aspect. The photograph is objective but like any other kind of illustration it is interpreted by the viewer. While both the photograph and the representational drawing may help the reader, therefore, they may also mislead him, and a diagram may serve your purpose better.

Diagrams

If you draw a diagram to scale, the scale should be marked in metric units. Such a diagram (or plan) conveys information more accurately than a photograph or drawing of the same thing. A map is a special type of diagram but there are difficulties in representing a globe to scale on a flat surface. On a map, north should be at the top of the page and should be indicated by an arrowhead.

Diagrams are also used in the presentation of data or statistics in graphs, histograms, bar charts and pie charts (Figs 9 and 10).

In diagrams which are not to scale, each line is not intended as an accurate record of an observation. The diagram as a whole may provide a summary of observations (Figs 11B and 12). It may help in presenting a new idea or in making comparisons (Fig. 4), in representing the order of events in a process (Fig. 15) or in showing the arrangement of equipment for an experiment (as in a circuit diagram).

The art of illustration

You are concerned with accuracy, not with artistic effect, but you should consider how best to present information in an illustration so that the reader is affected in the right way.

Balance, which makes a drawing appeal to the eye, is achieved only if you consider how the drawings or the parts of a drawing, and the labelling, are to be arranged on the page. Compose each drawing so that information is conveyed effectively, and use labelling to help the reader. If a drawing has several parts, use letters or arrows to guide the reader.

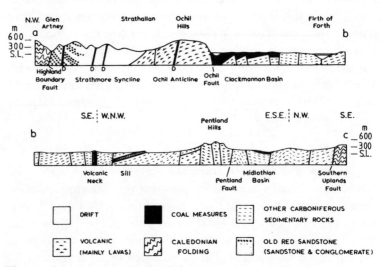

Fig. 12. *Diagram: a summary of observations.*
From Sissons, J. B. (1976) *Scotland*, Methuen, London.

So that your message is clear, do not clutter an illustration with too much information. If a graph or a drawing has too many lines, so that nothing stands out, the reader may have difficulty in distinguishing what is essential to your argument. Only you can decide what to leave out in the interests of clarity. An illustration should concentrate attention. For maximum impact the drawing and the message must be clear and simple, and the most effective illustration conveys one idea.

One way to prevent a drawing from being cluttered is to use two or more drawings instead of one. On a blackboard information is presented a little at a time and the teacher maintains contact with the class as an argument is developed or as a diagram is constructed. An artist uses the same technique in a strip cartoon. The subject is represented simply and in each drawing, with its caption, the message is clearly expressed. This technique can also be used in presenting science (Fig. 7), and is taken further in the preparation of graphs – in which each point represents an observation or is a summary of observations.

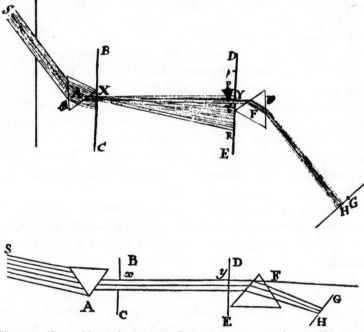

Fig. 13. *Above: Newton's diagram. Below: inaccurate redrawing by a later author.*
Both diagrams from Lohne, J. A. (1967) The increasing corruption of Newton's diagrams, *History of Science*, 6, p. 73.

Another way to prevent a drawing from being too cluttered is to use more of the page. Some subjects can be displayed effectively by using an explosion technique which helps to show how components fit together to form a more complex whole (Fig. 11C). The same kind of subject may lend itself to a cut-away technique – with superficial structures shown in their correct position but cut away to show deeper lying parts (also Fig. 11C).

In the interests of accuracy and clarity, each illustration should be planned to go with the text. Illustrations prepared for another purpose may save the writer time but anything irrelevant may confuse the reader. The uncritical re-drawing of illustrations prepared by others may also result in the perpetuation of errors; and new mistakes may be made because of lack of care or lack of understanding (Fig. 13). See also *Copyright*, p. 156.

Ink drawings for a report
Scientists who are convinced of the value of illustrations, as an aid to communication, are likely to accept that time should profitably be devoted to their planning and preparation.

An artist studies and interprets and then draws what seems to be important. The drawing conveys the artist's understanding of the subject. Because of this, it is best if you can illustrate your own reports. *You* know what to include, what to omit, and what labelling is needed. At least, you should provide the artist with as good a sketch as you can manage.

The house rules, or the instructions to authors, issued by the editors of journals and by publishers, should be consulted before ink drawings are started, so that you know about any special requirements.

Dimensions
Prepare each illustration so that it fits upright on the page and leave space for the legend. In a journal with a two-column format, the illustration may be the width of the column or that of the printed page (type area only).

When the size of a drawing has been decided you should use

a larger sheet of paper so that there are margins of about 40 mm. When the illustration is also to be used in the preparation of a slide for projection, its width will depend upon the width of a page or column in the report but the proportions of the drawing, height:width, must be 3:4 (4:3 is also suitable for a 5 × 5 cm slide).

If the drawings are to be reduced photographically for a report, or by the printer in the preparation of a block, they should be prepared so that they are twice the required dimensions (for reduction by one half) with lines twice as thick (0.3 mm for the axes and grid lines of a graph; and up to 1 mm for other lines on a graph). All letters, numbers and symbols must also be twice the required size, with lines twice as thick (except that some publishers require all the letters and numbers in pencil). Lettering with capitals 4 mm high is large enough for most purposes. Letters should be spaced so that there is a clear though narrow gap between them; and the space between words should be the width of a lower case 'm'. Lines of lettering should be well spaced. The space between ruled lines should be at least 4 mm. Every line in a drawing must be thick enough to show clearly after reduction, and any line that will not show should be erased. To facilitate the reading of numerical values, the graduations on the scales of a graph should be marked, for example, at intervals of 20 mm so that after reduction by half they are 10 mm apart.

Preparing a large drawing (for photographic reduction) encourages bold work, with large pens, on a large sheet of paper, and helps in the inclusion of detail. Small imperfections of line are less obvious after reduction and a good drawing then looks even better. However, reduction will not make an untidy drawing look neat. Neatness of line is essential in the drawing if every part and relationship is to be clear after reduction.

The reduction required should be shown in soft black or blue pencil on the margin of the illustration and a line should be drawn around these instructions to indicate that the words are not to be included in the finished illustration. The instruction *reduce by one half* will give a final size one half the dimensions of the original (a quarter of the original area).

If possible, the illustrations for one paper should be drawn with the same pens, on pages of the same size (say A4 = 210 × 297 mm), for reduction by the same amount; and when appropriate a number of drawings should be grouped together as one illustration (see Figs 10 and 11). The parts of an illustration should be designated by upper and lower case letters according to house rules. Grouping illustrations reduces costs of production and gives uniform lines and letters.

Drawing

Your final drawings should be in waterproof black India drawing ink on photographically white Bristol board, on blue-lined graph paper, on blue tracing linen, or on good quality tracing paper (110 g/m²).

A neat and even appearance is obtained if you work on the whole drawing rather than completing one part and then moving on to the next. Try to draw the whole of each line in one stroke of your pen. Draw straight lines with a ruler and curved lines with a compass, French curves or a flexible ruler. Use either complete, broken or dashed lines, or different symbols, to distinguish different curves on a graph.

Many publishers ask for the letters, numbers and labelling lines to be in pencil (so that these can be added in the house style). When you have to do the lettering in ink, use stencils and special pens for letters, numbers or symbols (or use transfers).

Graphs may be prepared on blue-lined graph paper and then photographed or traced. While grid lines are essential for the preparation of a graph they are not usually necessary for its interpretation (and the blue lines will not appear in the photograph). If any lines are required, therefore, these must be added in black ink.

The points on a graph should be indicated by symbols; and when a graph is intended for publication these should be the symbols that are available to the printer (see the journal's *Instructions to Authors* or the house rules) so that the same symbols can be used in the legend, e.g.:

○   ●   □   ■   △   ▲

If other symbols are used a key should be included as part of the figure (not in the legend).

The same symbols and line forms should not be used on two curves in one graph if the points may be confused, but the same symbols should be used for the same quantities throughout a paper.

Each axis of a graph should be labelled, parallel to the axis and on its outside. Numbers on the axes should also be outside the graph next to small bars which project on the inside.

Writing the legend (*caption*)
Because more people will look at the figures than will read the text, it should be possible to understand each illustration without reference to the text. The legend should be complete, clear and concise.

A statement in the legend should indicate whether the symbols on the graph are records of observations or arithmetic means. If vertical lines are drawn through the symbol, above and below the mean, the legend must indicate whether these show the standard error ($S\bar{x}$). the 5 per cent fiducial limits of error ($\pm 1.96\ S\bar{x}$) or the range.

Do not clutter your illustration with information that can be put in the legend, but a scale on the figure is better than a statement of magnification in the legend, and a key to any shading or symbols should be included on the figure rather than in the legend.

When an illustration is used to inform, it must be a correct record of an observation or a summary of observations; and the legend (in the present tense) must be a factual explanation. But when a diagram is used to convey ideas, this must be made clear in the legend.

Any help, the source of data, and the source of any illustration that is not original, should be acknowledged: see legends to Figs 7, 9 and 12, and *Copyright*, p. 156.

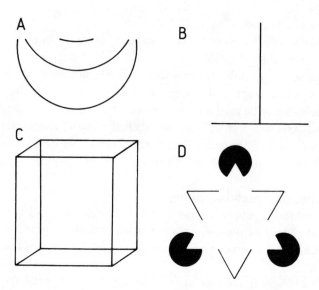

Fig. 14. *Optical illusions: (A) arcs of equal radius; (B) horizontal and vertical lines of equal length; (C) a cube which may be interpreted differently at different times; and (D) an illusion of a whiter triangle.*

Completed illustrations

Store your illustrations in an envelope or between sheets of clean paper, and with a cardboard sheet to prevent bending, so that they do not get soiled or creased.

Check the figures and legends and get someone else to check them, to ensure that they do serve their purpose. People will believe what they see. Subconsciously a drawing presents difficulties for the artist and for the viewer (Fig. 14). The artist has to represent the subject in two dimensions and the viewer has to interpret the drawing so that he sees the subject as if it were in three dimensions. If possible your drawings should be simple, with clear lines, but additional artwork or labelling may be needed to facilitate interpretation.

Things to check in the illustrations

1 Check every illustration against your original artwork.
2 Check for clarity, accuracy and neatness of line.
3 Does the numbering of the figures correspond with the numbers in the text; and are other things referred to in the text included?
4 Are the letters, words and abbreviations on the figures consistent with those used in the text? Is the labelling clear?
5 Is any figure cluttered with too much information?
6 If the figure is to scale, is this scale marked on the illustration?
7 Are the units of measurement clearly marked on all axes and scales?
8 Are all symbols sufficiently explained?
9 Should any photograph be replaced by a line drawing?
10 If the illustrations are to be published, check that the information required by the printer is written in the margin or on the reverse in soft black or blue pencil: author's name, title of paper, number of figure, reduction required. For some journals all the illustrations are numbered consecutively; but in others the figures and plates are numbered separately with Arabic numerals for the figures and Roman numerals for the plates (photographs). Letters should be used to designate the parts of a figure (Fig. 11A, B and C, for example).
11 Check that the information in a table is not duplicated in an illustration.
12 Check that each table and illustration will fit upright on the page in the space available, and check that there will be space for the legend.

I I

Reading

We learn many things by discovery but most of what we know comes from conversation or reading. Discoveries are made against a background of existing knowledge which forms part of the opportunities of place and time.

Reading may save you the fruitless labour of seeking, by observation and experiment, information which is already in the literature; but do not be convinced too easily that something you wish to investigate has already been studied exhaustively. What is written is not necessarily true and is seldom the whole truth. Even accurate observations may be incomplete and changes in technique or in the design of experiments may lead you to new observations, to a different interpretation, and to new lines of enquiry.

Some young scientists may have too much respect for the printed word, but experienced scientists may agree that: 'It is that which we do know which is the greatest hindrance to our learning, not that which we do not know' (Claude Bernard). 'The advantage of a certain amount of ignorance is that it keeps you from knowing why what you have just observed could not have happened' (Sir Frederick Gowland Hopkins). Reading, before you investigate, may direct your mind along well worn tracks and away from a fresh approach to a problem.

Books are less often made use of as 'spectacles' to look at nature with, than as blinds to keep out its strong light and shifting scenery from weak eyes and indolent dispositions. The bookworm wraps himself up in his web of verbal generalities, and sees only the glimmering shadows of things reflected from the minds of others.

On the Ignorance of the Learned, William Hazlitt (1778–1830)

Relate the time you spend on your literature search and on reading to the amount of time available. It is better to work in ignorance of what has been done than to spend so long in searching the literature that you have no time for observations and experiments. Also, be prepared to observe and experiment as the opportunity arises. Your first observations will make possible a more informed reading of other people's work and serve, if nothing more, as a trial run.

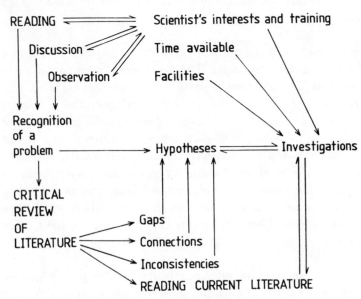

Fig. 15. *Reading as part of a scientific investigation.*

Other people's work

Observations by other people may complement your own or suggest new ways of looking at a problem and new lines of investigation. You may also link ideas previously unconnected in the minds of others, or improve your own procedure, after you have considered how others have tackled similar problems.

Relating new observations to earlier work should also lead you to a deeper understanding of your problem. While you may work alone, science is a universal language and most new discoveries are the result not only of observation and interpretation but also of communication between specialists.

With so many scientists, and with more books and journals published each year, it is difficult to know what other people have written on any subject. Starting with recent publications, you will find references to related papers in a wider field. It is unwise to restrict your reading to those aspects of a problem which are of immediate interest, because new ideas may come from unexpected sources.

Encyclopaedias are a good starting point for anyone coming new to a subject (Grogan, 1973). The best known, most authoritative and most detailed encyclopaedias include the *Encyclopaedia Britannica*, *Le Grand Larouse Encyclopédique*, and *Chambers Encyclopaedia*. In the multi-volume *Encyclopaedia of Science and Technology* (McGraw-Hill) each article is written so that it can be understood by non-specialists. Van Nostrand's *Scientific Encyclopaedia* is in one volume. The *Concise Encyclopaedia of Electrical Engineering* (Newnes, London), is an attempt to provide information for electrical engineers in those fields in which they are not experts. There are similar encyclopaedias on most other subjects.

Handbooks are concise reference books for day-to-day use by specialists. They provide information on one branch of science (e.g. *Handbook of Chemistry and Physics*: Chemical Rubber Co., Cleveland) or on one topic (e.g. R. A. Fisher and F. Yates *Statistical Tables for Biological, Agricultural and Medical Sciences* (Oliver and Boyd, Edinburgh).

The Complete Plain Words (Gowers, 1973) is a handbook for those who use words as tools of their trade, and *Usage and*

Abusage (Partridge, 1965), a book on English usage – published first in the United States – includes notes on American usage (see also Fowler, 1974). The Royal Society of London publishes *General Notes on the Preparation of Scientific Papers for Publication*, and more detailed style manuals and guides are published, for example, by the CSIRO in Australia and by the American Medical Association in the U.S.A.

Standards concerned with different aspects of scientific writing are:

American National Standards Institute (ANSI)

Y1.1	Abbreviations for use on drawings and in text.
Y15.1	Illustrations for publication and projection
Z39.4	Indexes, Basic criteria for (see ISO 999 below)
Z39.5	Abbreviation of titles of periodicals (ISO 4; ISO 833; BS4148)
Z39.14	Writing abstracts
Z39.16	Preparation of scientific papers for written or oral presentation
Z39.18	Format and production of scientific and technical reports
Z39.22	Proof corrections

British Standards

BS 350	Conversion factors and tables (in two parts)
BS 1219M	Preparation of mathematical copy and correction of proofs
BS 1629	Bibliographical references
BS 1749	Alphabetical arrangement and the filing order of numerals and symbols
BS 1991	Letter symbols, signs and abbreviations
BS 3700	Preparation of indexes to books, periodicals, etc.
BS 3763	The International System of units (SI)
BS 4148	The abbreviation of titles of periodicals (agrees with ISO 833 and ANSI Z39.5)

BS 4811   Presentation of research and development reports
BS 4821   Presentation of theses
BS 5261   Copy preparation and proof correction (2 parts)
BS 5261C Marks for copy preparation and proof correction
BS 5555   SI units and recommendations for the use of their
          multiples and certain other units (identical with
          ISO 1000)
PD 5686   The use of SI units (agrees with ISO 1000)

International Organization for Standards (ISO)

ISO 4   International code for the abbreviation of titles of
           periodicals
ISO 31/0   General principles concerning quantities, units and
           symbols
ISO 31/1   Basic quantities and units of the SI and quantities
           and units of space and time
ISO 214   Abstracts and synopses
ISO 215   Presentation of contributions to periodicals
ISO 690   Bibliographical references
ISO 833   International list of periodical title word abbrevia-
           tions (with which BS4148 and ANSI Z39.5 agree)
ISO 999   Index of a publication
ISO 1000   SI Units (identical with BS 5555)
ISO 2145   Numbering of divisions and subdivisions in written
           documents

Directories provide names and addresses and sometimes other
information, and cover many subjects including the names of
scientists, trades, industries and organizations (Harvey, 1969). You
should know of such lists of authors and titles as *Books in Print*,
Medical Books in Print, and *Scientific and Technical Books in
Print*.

Books It is not possible to keep all the books on one subject together
on the shelves of a library. If you wish to know which books are

available on any subject, first look at the subject index in your library. Then, to find which books are in the library, consult the appropriate classification numbers in the subject catalogue. Here you should find a card for each book. The book number on this card indicates where the book, with the same number, is to be found on the shelves.

The easiest way to find if a particular book is in your library is to use the alphabetical catalogue. Most books are listed according to the name of the author or editor but some are listed by the name of the organization, government department or society which produced the book. Each index card includes, in addition to this name, details of the book and the book number.

Reviews of the literature on a particular aspect of science are usually published in books or journals which specialize in review articles; but some are published in primary research journals. In a review, credit should be given to all those whose published work has advanced the subject, or whose work would have done so had it not been overlooked (UNESCO, 1968). All relevant published information should be assembled, analysed and discussed. A good review, therefore, is a good starting point in a literature survey. However, reviews may say nothing about the methods used and references to earlier work are necessarily brief and some are misleading. Read the original paper therefore if the work referred to has a close bearing on your own.

Specialist journals are for the publication of the results of original research. In looking at current journals a scientist sees interesting papers soon after they are published, and these papers include references to related papers which may also be of interest. However, it is not possible for a library to subscribe to all journals (Fig. 16), nor for any specialist to read all those that might be of interest. Another problem is that papers are not necessarily published in the most appropriate journal. As a result, you will not see all the papers that you might like to read.

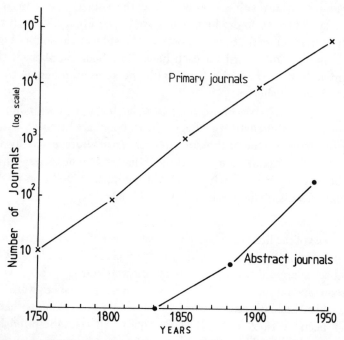

Fig. 16. *Number of journals founded 1750 to 1950, including those that are no longer published. Logarithmic scale on y axis.*
Estimates from de Sola Price, D. (1975) *Science Since Babylon,* Yale University Press.

Abstracting journals include abstracts of published papers, with a full bibliographic reference to each paper. This information will help you to trace relevant papers but abstracts lack detail and, like review articles and books, they also lack the enthusiasm of the original. They should be regarded as an introduction to the original paper not as a substitute.

Indexes are published as part of many journals but some have no index and others have one that is incomplete and therefore misleading.

Indexing systems covering many current journals include: MEDLARS II a computerized version of the *Index Medicus*; the *Applied Science and Technology Index*; *Pandex – current index to scientific and technical literature*; and *Bioresearch Index*. The Royal Society's *Catalogue of Scientific Papers* is an author index to the papers published in 1500 journals in the nineteenth century.

The *Science Citation Index* provides an index to more than 2000 journals published in any year (source journals) and also to earlier papers cited during the year. A literature search based on this index can start with a reference or an author's name, and this leads to other authors who have cited this paper and therefore to related literature. From this index a scientist may also determine whether or not his own work has been applied, extended or criticized.

Current Contents, a weekly, is published in six parts: Agriculture, Biology and Environmental Sciences; Behavioural, Social and Educational Sciences; Clinical Practice; Engineering and Technology; Life Sciences; and the Physical and Chemical Sciences. Each issue comprises the contents pages of current journals and a list of the authors' addresses.

Making notes as you read
Many readers spend time making detailed notes. If you do this as a habit, consider whether or not your time could be better spent. Students should buy up-to-date books on each aspect of their courses. They should then learn in two ways: by reading each book several times; and by preparing notes on subjects of particular interest.

In preparing notes, decide what information you require, then identify relevant passages in books and articles, and then relate the information and ideas from your reading to what you have heard in lectures and to your own observations. Your notes should include key words and phrases, headings and sub-headings, concise summaries and simple diagrams and drawings (but not long passages copied from text-books since these can be consulted again and again). Brief notes are an aid to study. They will be similar to – and will be most useful in the preparation of – a topic outline for an essay or report.

Keep a record of everything that you read. Note the author's name and other details both as a heading to your notes and on an index card (in the standard form used in a bibliography; see p. 140). Then keep your cards in alphabetical order for quick reference. You may use them later in tracing the same publication or in citing the references in a bibliography. If the card is large enough (203 × 126 mm), there is space for a summary, for notes or quotations, or for a reference to notes, a photocopy or a reprint kept in another place, or for the code number of a book in a local library.

How to read

Read to get background information and in search of information on specific points. You do not need to read the whole of every book or paper that you consult. Some books are written as reference books but even those which can be read as a whole can also be read in part to get just the information that you need at the time. This is a good way to start reading about a subject, since you will remember best those things that you find most interesting.

It is best to start with recent publications on any subject to find the present position and to be guided by your special interests to earlier literature. In deciding how far back to go, an important consideration is the amount of time available. Most people enjoy reading the work of others who are interested in the same subject as themselves and observations recorded, however long ago, are as correct today as when they were written.

When you have decided what to read, be it the title, the summary, the introduction, or the whole text, remember that an effort is required of the reader as well as of the writer. Scientists should practise an economy of expression, and should write carefully so that readers may grasp the meaning quickly; but this can be so only if each word is read.

Read carefully to make sure that you take the intended meaning.

Read critically, as a stimulus to thinking. Weigh the words and consider the evidence and arguments. These questions should be in your mind: What is being said? Are the underlying assumptions correct? Are the statements supported by sufficient evidence? What are the implications of the work? Is there a better explanation of the evidence presented?

Reading the prose of other scientists is the student's introduction to the conventions of scientific writing. In reading, scientists learn more about science and from those who write well they learn how information and ideas may be presented clearly and concisely and in an interesting way.

Writing a book review

Some working scientists write reviews of books, and a book review may also be undertaken by students as an exercise in comprehension and criticism. In preparing a book review, reading and writing should be associated closely.

The length of the review may be decided by the editor. If the review is too long it may be reduced by the editor. He may do this by deleting sentences at the end; so the most important things must be written first and the least important things last.

The reader requires: title of book (and sub-title); name(s) of author(s) or editor(s) from the title page; date of publication; number of edition (unless the first); name of publisher; place of publication; total number of pages (including preliminary pages); number of tables, figures and plates; and price of hardback and of paperback.

Readers of a book review have been attracted by the title. They want a brief guide to the book and an evaluation to help them to decide whether or not to read the book. Answer the following questions. What is the book about? Has it any special features? How is the subject treated? What prior knowledge is assumed? Who is the author writing for? Is the treatment comprehensive? Is the writing interesting and easy to read? Are the illustrations effective? Is the book well organized? Will the reader, for whom the book is intended, find the book useful? How does the book compare with similar books (if there are any) or with this author's earlier works?

Reviewers who have never written a book are unlikely to appreciate the writer's difficulties. Perhaps this is why some reviewers seem to be looking for the perfect book. While you may choose to draw attention to errors, if these indicate that the author is not as knowledgeable as he should be, it is not your task to list every minor fault.

The purpose of a book review is not to show that the reviewer is (or is not) clever and witty, but to inform the reader. Yet some book reviews tell the reader more about the reviewer than about the book. You should not try to demonstrate that you could have written a better book; nor try to score at the author's expense; and there is no place for sarcasm or for rhetorical questions in a review of a book about science.

12

Reports and theses

The parts of a research report

The use of the commonly accepted headings will help you to organize your work and will help readers to know where to look for the information they require.

The editors of some journals specify the headings to be used unless there is some overriding reason for not doing so.

Introduction	What did you do? Why?
Materials and Methods	How did you do it?
Results	What did you find?
Discussion	Your interpretation of your results.
Summary	Statement of main findings.
Acknowledgements	Who helped?
References	Details of references cited.

Some employers have house rules which must be followed in the preparation of internal reports. Otherwise, there are standards for the presentation of research and development reports (BS 4811 and ANSI Z39.18): see Table 18.

The front cover

The front cover of a research and development report should include: a serial number; the title in full (including any sub-title); the author(s) name(s); the name of the organization (and of the division of the organization) responsible for the report, and its full postal address; the date when ready for reproduction; the price (and sales point if different from the organization responsible for the report); the security classification, if any. Any special notices

Table 18. Arrangement of a research and development report

Front cover   ⎫
Title page   ⎬   *or* Report documentation page (ANSI)
Summary (abstract)   ⎭
Preface (not usually needed)
Table of Contents (needed for all except short reports)
Introduction
Theory (additional to or alternative to next section)
Experimental procedure and results (with sub-headings)
Discussion
* Conclusions (must be precise, orderly, clear and concise)
* Recommendations (arising directly from the conclusions)
Acknowledgements
List of references
Appendices
   Tables   ⎫
   Illustrations ⎬   if not included in the
   Graphs   ⎭   main body of the text
   Literature survey (if necessary)
   Bibliography (supplementary to list of references cited in text)
   Glossary (if necessary)
   List of abbreviations, signs and symbols (or *after* Contents: ANSI)
Index (if necessary)
Distribution list (if required by the sponsor or house rules)
Document control sheet
Abstract card(s)
Back cover

---

* Alternatively, the conclusions and recommendations may be placed
immediately after the introduction.

required by the sponsoring organization should be on the inside
of the front cover.

In a bound report or thesis, after the front cover, the first sheet
is blank, and the next (the half title) has only the title. The next
page is the title page.

The title page
The title is the first and most important part of any communi-

cation. Just as you read quickly through the headlines of your newspaper – to see if there is anything worth reading today – so scientists read the *Contents* page of a journal. On reading the title, they decide to look at the *Summary* or at the *Introduction*, or they stop reading. It is worth giving a lot of thought to the choice of a good title to ensure that it attracts the attention of all those who might wish to read part or the whole of the report.

Remember that the title of an internal report should be useful to the many people who will see only the title – in a memorandum, a research report, or in a list of references. The title of a published report should be useful to people who see only the title – in an index or abstract or in a list of references in another paper – as well as to those who have the whole paper to examine.

The title should be brief but unambiguous, and it should give a clear indication of the subject and scope of the work. Some journals print lists of key words (words used in indices) and key words that might contribute to information retrieval should be included in your title. The title used at the start should be reconsidered at the end. Check that it is sufficiently direct and informative. Delete superfluous words (for example: Aspects of . . .; Neglected effects of . . .; Studies on . . . Part . . .).

Below the title, list the name(s) of the author(s) of the report and the name(s) of the institution(s) in which the work was done (the full postal address). Date the report and give it a serial number. If it is different from the place at which the work was done, give the author's present address as a footnote.

If more than one person has contributed to the work, the authors' names should be in alphabetical order, or in an order which reflects each person's contribution, or in an order determined by house rules or national custom.

For a printed report, write the following information on the title page, and draw a circle around it to indicate that it is not to be printed: a note of the number of folios (pages of typescript, tables and illustrations or other copy); a name and address for the editor to use in correspondence; and a short running title for the top of each printed page.

The Table of contents
The way through your report should be clearly sign-posted by
main headings and appropriate sub-headings, and above a certain
size a table of contents is also needed. Decide whether or not a
Contents page would help the reader, and if there is any doubt
this should be included.

On the *Contents* page all the principal headings, and perhaps
also the sub-headings (indented), should be listed in the order in
which they occur in the report. In the final draft of the report,
therefore, the pages must be numbered. Page numbers also
facilitate cross reference.

The Introduction
Your report on an investigation should begin with a clear statement
of the problem (if there is a problem) indicating the scope of the
work, what the report is about, and why the work was undertaken. Include a brief reference to any preliminary note and to
other relevant investigations, by yourself or by others, to show
how the work to be reported is based upon earlier work. Mention
any new approach, any limitations, and any assumptions upon
which your work is based.

In many journals the *Summary* is printed before the *Introduction*
and it should not be necessary to repeat things in the *Introduction*
which should properly be in the *Summary*.

A clear, concise and interesting beginning may encourage the
reader to continue reading. Write, as in the *Summary* and in your
conclusions and recommendations, in straightforward non-
technical language, so that your introduction can be understood
by all the people for whom the report is intended; even if the
Methods, *Results* and *Discussion* sections can be understood only
by specialists.

The Materials and methods
Include enough detail to ensure that if the investigation is repeated
by someone else, with experience in the same field, similar data
could be obtained.

Clearly, er, I mean, obviously...

Fig. 17. *For maximum impact each illustration should be clear and simple.*

1 List the equipment used and draw anything that requires description (unless this is very simple).

2 State the conditions of the experiment and the procedure, with any precautions necessary to ensure accuracy and safety. However, when several experiments are reported some details may fit better in the appropriate parts of the *Results* section.

3 Write the stages in any new procedure in the right order and describe in detail any new technique, or modifications of an established technique.

4 If necessary, refer to preliminary experiments and to any consequent changes in technique. Describe your controls adequately.

5 Include information on the purity and structure of the materials used, and on the source of the material and the method of preparation.

6 Use systematic chemical names or the pharmacological names recognized in the country in which your report is to appear. Give the internationally accepted name of the thing studied and refer to any factors which influenced your choice of material. Describe your techniques or refer to an earlier report in which these are described.

First, have something to say!

Fig. 18. *If you think of your report as part of your investigation, not as a duty to be undertaken when your work is otherwise complete, questions such as when to start writing or which part to write first do not arise.*

The Results

This section, written in the past tense, should provide a factual statement of what you observed supported by any statistics, tables or graphs derived from your analysis of the data recorded during your investigation. Any other diagrams should normally be included with the results as an aid to their interpretation.

If necessary, as in a thesis, the original data can be included in tables in an *Appendix*. The tables in the *Results* section should be summaries.

Describe representative successful experiments in detail; and it may be helpful to mention briefly the unsuccessful experiments and wrong turnings which are part of every investigation.

Present your results in a logical order (not usually the order in which you did the work). Remember that this is not the place for your comments.

The Discussion

This should be an objective consideration of the results given in the previous section and should lead naturally to your main conclusions.

Refer to any further light cast upon the problems raised in the *Introduction* and say how your work fits into the background of previous investigations; but claim no more than you can substantiate.

References to what you did should be written in the past tense to emphasize that you are commenting on the work reported; but write statements of fact in the present tense.

You may be tempted to include much information sifted from the work of other people. However, many of these sources can be adequately covered by a few words followed by a reference. Other things that seemed relevant when you made your notes may have no place in your final report. The references cited should provide essential background for which you have no space in the report or should be needed for the development of your argument.

Most scientists have been misquoted or misunderstood. They may be pleasantly surprised when they are not. When reading, therefore, use abstracts and reviews but go to the original to find what other people did, how they did it, and what they concluded. When writing, take care to make your meaning clear. Anything ambiguous is likely to be misunderstood.

Take care, when quoting from the work of others, to check that the words and punctuation used are copied correctly. When you summarize other people's work try to preserve their meaning. This is not easy and when the reference is out of context readers may take a different meaning from the original.

The Summary
The summary can be written only when the report is otherwise complete. It is placed immediately after the title (where it may be called a synopsis or abstract) or at the end of the *Discussion*.

In a published paper the summary (with the title) should be suitable for use as an abstract by information services (UNESCO, 1968). If you write a summary that is too long, someone else will shorten it and they may leave out things that you consider important. Remember, also, that if someone refers to your paper in a book or review they may reduce your conclusions to one sentence or to a phrase. Can you provide a suitable sentence in your summary?

Great care is needed in the preparation of the summary because, after the title, this is all that most readers will see. It must be complete, interesting and informative without reference to the rest of the report, except that information given in the title should not be repeated.

Your summary should be as short as possible (usually less than 200 words) but everything new and everything you particularly want people to know must be mentioned. The problem should be stated; and the main findings and conclusions should be included in the same order as in the report. No table numbers, figure numbers, references or citations should be included. There should be no information, ideas or claims other than those in the report.

The treatment of the subject may be indicated by such words as preliminary, detailed, theoretical and experimental. When experiments are reported the methods used should be mentioned. For new methods, the basic principles, range of operation, and degree of accuracy should be given. Statements of conclusions and inferences should be accompanied by an indication of their range of validity. If the journal publishes more than one kind of article, the category to which your paper belongs should be mentioned (preliminary note, original paper, or review).

The summary should be in the third person, in complete sentences, and in words that will be understood by everyone for whom the report is intended. Ask someone who has not read your report to read and comment on your summary. See also *Standards*, p. 123.

The Acknowledgements

Any acknowledgements should be typed on a separate sheet. In this section refer to the source of finance, and to anyone who helped either in the work (with materials, assistance or advice) or in the preparation of the report. Some writers add that they take responsibility for the final arrangement, the opinions expressed, and for any remaining mistakes. In scientific writing this truism should be omitted but some statement may be required, for example by a firm or government department, to indicate that the views expressed are not necessarily officially endorsed. This is another occasion when the house rules must be followed.

Allow anyone mentioned in the *Acknowledgements* to read what you have written about them so that they have the opportunity to comment. Do not refer to unpublished work (personal communications) or use copyright material unless you have written permission (see p. 156). The source of copyright material must be acknowledged.

In a report on an original scientific investigation the question of copyright does not usually arise, but when quoting from someone else's work the quotation must be accurate and the source of the quotation must be given. Particular care is needed in checking the accuracy of all references, including the spelling of proper names, since a reference is both an acknowledgement of the work of others and a source of further information for the reader.

The List of references

The way in which references are cited in the text and listed at the end of a paper must be in accordance with the house rules or the practice of the journal in which the paper is published. The system adopted in this book is widely used.

Authors are named in the text by their surname only (and the date of publication is given in parenthesis). When reference is made to more than one paper by the same author, published in the same year, these should be distinguished by suffix letters a, b, c, immediately after the date. If a paper has two authors both names are given. If there are more than two, all names are listed the first time the paper is cited, and for later citations the name of the first author is followed by *et al.*

All the references cited in the text should be included in the list of references, but no others. Reference to unpublished work or to papers which are in preparation should not be included. The references are listed in alphabetical order (see BS 1749: *Alphabetical arrangement*) according to the surname of the author (or editor), or that of the first of a number of authors (or editors). An editor (ed.) or translator (trans.) is indicated by an abbreviation after the name.

1 The surname(s) and initials of the author(s) or editor(s). The year of publication may follow in parenthesis, or the year may be placed later (ISO 690) as indicated in 2 and 3 below.

2 The titles of books are underlined in typescript (and printed in italics) and initial capitals are used for most words. The title is followed by the number of the edition (except for the first edition), the number of volumes (e.g. 2 vols.) or the volume number (in arabic numerals, underlined with a wavy line in typescript and printed in bold), but not the abbreviation vol., the place of publication, the name of the publisher, the year of publication, and either the number of the page (p.   ) or pages (   ) referred to, or the number of pages in the book (pp.).

3 The titles of papers are not underlined and initial capitals are used only for words that normally require them. The title is followed by the name of the journal (underlined in type-

script), the volume number (underlined with a wavy line) but not the abbreviation vol., the issue number in parentheses, the date of issue, and the numbers of the first and last pages of the paper (without the abbreviation pp.). If the names of journals are abbreviated, the abbreviations should be in accordance with the recommendations of the *World List of Scientific Periodicals* (Butterworths, London) or with International Standards (p. 124). However, in research and development reports the titles of periodicals should not be abbreviated (BS 4811).

4 The title of a chapter contributed to a book is followed by the word *in*, and this by a colon, the name of the editor(s), and the abbreviation (ed.) or (eds.), and other details as for a book, and then the first and last pages of the paper.

A complete literature citation is necessary for information retrieval. If a citation is incomplete other people (readers and librarians) must waste time in trying to find the missing details.

A bibliography differs from a list of references in that it is not confined to publications cited in the text, and annotations may be included. A note, after the heading, should state the principles on which the bibliography has been compiled. Recommendations for bibliographical references are the subject of the International Standard ISO 690 (and of BS 1629).

Project reports and theses

Theses

The word thesis means a statement, proposition or position which a person advances and is prepared to maintain. The word is also used as a synonym for a dissertation: a written presentation of a subject, a contribution to knowledge, usually prepared by a candidate for a higher degree.

The purpose of a thesis is to train the mind of the writer and to show how far it has been trained. The writer of a thesis, after years of thought and study, must have made the subject his own, and his writing must convey the ideas and understanding that come from observation and reflection.

The thesis for a Master's degree is based upon a training in the problems and methods of scientific investigation: upon *independent* research. The thesis for a Doctorate is based upon independent *original* research: upon an investigation in which the frontiers of knowledge have been explored and extended.

The limited purpose of a thesis is to contribute to the solution of a problem. Great care is needed, therefore, in selecting a problem to which a contribution may be made in the time available.

The parts of a thesis are the same as for other scientific reports (see advice on pp. 131–41). However, because it has to satisfy an examination requirement, the thesis has a restricted readership: it is written for the few specialists who will judge its merit.

The title page should include the full title (and any sub-title); the full name of the candidate; the qualification for which the thesis is submitted; the name of the institution to which the thesis is submitted; the department, faculty or organization in which the research was conducted; and the month and year of submission.

The title page should be followed by a table of contents and a list of tables and illustrations. The *Acknowledgements* should include a declaration in which a note is made of any material in the thesis which has been used before, and of the author's part in any joint work included in the thesis. The *Summary*, of about 300 words, should state the nature and scope of the research and of the contribution made to knowledge of the subject. A brief summary of the method of investigation may be appropriate, an outline of the major divisions or principal arguments of the work, and a summary of any conclusions.

After a short introduction, a concise critical literature survey of relevant previous work may be required (as a separate section). This gives perspective to the work and shows existing knowledge as the basis of further discoveries. However, candidates should refer only to those things that they can discuss intelligently in their oral examination.

The *Materials and Methods* and *Results* should show the approach to the problem, and what has been added; and in the *Discussion* the results are interpreted and discussed in relation to previous work. The references cited must make clear the writer's understanding of relevant literature in the general field, and of all references directly concerned with the problem. In all its parts the thesis is a measure of scholarship and industry as well as of research ability.

The thesis must meet the detailed requirements of the examining body to which it is to be submitted (see also BS 4821 on the *Presentation of theses*) and it should normally be similar in form and arrangement to theses previously submitted for the same qualification in the same Department. However, many theses are too long and contain more tables and illustrations than are necessary. Subject to the supervisor's agreement, the body of the thesis (the part following the critical survey of previous work) may be prepared in a form which is suitable for immediate submission to an appropriate journal, as one or more research papers. If the original numerical data are included they should be in an Appendix.

Project reports

These have some of the features of a thesis, when a project is completed as part of a training in research, and project work is part of many first degree courses.

The report of a project is more demanding than anything that a student has written previously. It refers to an investigation which has taken much time, and which may have involved different kinds of observations and experiments as well as a consideration of other people's work.

Different types of project will be reported in different ways. For example, a project which is based on a literature search and upon reading, with no supporting personal observations, will be written as an extended essay or review: see Chapter 5. On the other hand, a project in which the student's own observations are reported and related to relevant work by others should be written as a scientific paper with the same headings.

A project should provide scope for initiative, ingenuity and originality. The report should therefore indicate not only what was done but also the student's approach to the problems involved, to the interpretation of work by others, and, if appropriate, to the analysis and interpretation of new observations and results. Because of these things the written report, like the thesis, plays a major part in assessment, just as the published paper affects a scientist's reputation.

Project assessment

It is difficult to arrive at an objective assessment of extended practical work and projects, but easier to assess the quality of the written report and to make a subjective judgement based on the student's general attitude to the work.

It may be difficult for an examiner to decide how much of the report is the student's work and how much is the supervisor's. In planning a project the student should be guided, to ensure that the project is one that can be completed in the time available. Also, guidance should be given at the start of the project to ensure that the work begins well.

Different projects involve different methods of investigation: some require more background reading than others; and some provide more scope than do others for the student to show originality, initiative and ingenuity. These differences, which make objective assessment difficult, must be carefully considered if all students are to be treated fairly.

There is probably no one solution to these problems but it is possible to list things which should influence the final assessment. These cannot all be judged at the end of the work.

1 The student's approach to the project – his ability to define the problem clearly and to tackle any difficulties.
2 The thoroughness with which the work was tackled in relation to the time available and the logical planning of the procedure.
3 The accuracy with which information is recorded.
4 The student's ability to interpret work by others, or personal

observations, or to analyse numerical data (as appropriate) and to draw conclusions.

5 The student's ability to relate his findings to his knowledge of the subject derived from the work of others.

6 The presentation of the work as a clear, concise and well organized written report.

These aspects of the work might be considered of comparable importance and given equal weight in a marking scheme. But, whatever the method of assessment, each student must be told, before the project is started, how the project should be presented as a report and how the project as a whole will be assessed.

13

Preparing a report on an investigation

Preparing the manuscript

Prepare the *Contents* page first. This will help to remind you of the kind of information to be included in each part. Your list of headings and sub-headings should then be revised each time you add or delete anything, to help in the orderly arrangement of information and to prevent repetition.

Your first draft of the *Introduction* should include a clear statement of the problem. Write this before you start the investigation. Defining the problem will concentrate your attention on the purpose of your work. Write the *Materials and Methods* section as soon as you have standardized the procedure while details and difficulties are fresh in your mind. Prepare and revise the *Results* section as your work proceeds. The tables and illustrations are usually so important that you must prepare these first and then write the text. By writing a summary of each part of your investigation, as soon as it is complete, you present information while it is fresh in your mind and, if necessary, you can repeat the work or obtain more records while the equipment is at hand.

Notes on points to be included in the *Discussion* can be assembled throughout the work but your discussion of the results can be written only when all your results are available. Then, the *Introduction* should be revised and the *Summary* prepared.

The logical order in which information is presented in a scientific paper is not usually the sequence of discovery. But this logical order makes for easy reading and for an economy of words.

Write each heading and sub-heading, and each paragraph, on

a separate sheet of paper; and prepare each table and graph on a separate sheet. You can then keep information on each aspect of your work in the right place, incorporate new material, or revise the order of presentation. Unwanted material can be removed. In this way, while you are working, your draft remains an up-to-date progress report.

Write in ink, on one side only of wide-lined A4 paper. Leave wide margins and write on alternate lines. Keep a carbon copy of your manuscript, in another place, and correct the copy each time you revise the manuscript.

If you add anything to the early draft, as a footnote, incorporate this in the text when you next revise the page. A footnote may be needed on the title page; otherwise avoid footnotes unless they are essential in a table. Do not use footnotes for information that ought to be in the list of *References* or *Acknowledgements*.

Things to check in your manuscript
It is not possible to check the completed manuscript thoroughly by reading it through once or twice. It is better to check one thing at a time.

1 Is the title page complete (see p. 133)?
2 Does the title provide the best concise description of the contents of the report?
3 Is the *Contents* page still needed? If so, are the headings the same as those used in the report?
4 Is the use of headings and sub-headings consistent through-out the report; are the headings concise; and are all the headings and sub-headings used in planning the report still needed?
5 Is the purpose and scope of the report stated clearly and concisely in the *Introduction*?
6 Have you achieved your purpose and kept within the terms of reference?
7 Has anything essential been left out? Are all the reader's questions (see p. 37) answered? Are your conclusions clearly expressed?

8 Is each paragraph relevant, necessary and in its proper place? Are the paragraphs in each section in the most effective order? Is the connection between paragraphs clear? Cross out anything that is irrelevant.

9 Is each paragraph interesting? Is the topic clearly indicated and is everything in the paragraph relevant to the topic? Is the emphasis in the most effective place?

10 Are all arguments forcefully developed and taken directly to their logical conclusion, and is anything original emphasized sufficiently?

11 Is there an important point which could be more clearly expressed, or which should be put more forcefully in an illustration? Should any illustration be replaced by a few lines of text?

12 Does the report meet all the requirements of scientific writing (Chapter 4)? Is each statement accurate, based on sufficient evidence, free from contradictions, and free from errors of omission? Are there any words such as *many* or *a few* which can be replaced by numbers?

13 Are there any faults of logic or mistakes in spelling or grammar?

14 Is each sentence necessary? Does it repeat unintentionally something that has already been written?

15 Could the meaning of any sentence be better expressed? Are there unnecessary words?

16 Is each sentence easy to read? Does it sound well when read aloud, and is the emphasis in the most effective place?

17 Are any technical terms, symbols or abbreviations sufficiently explained?

18 Are all the words that are to be printed in italics underlined?

19 Are you consistent in spelling, and in the use of capitals, hyphens and quotation marks?

20 Are all the references accurate, especially the spelling of names? Do the dates in the list of references (on your index cards) agree with those given in the text?

21 English is a language of international communication. If your

report is for a wide readership, or for readers with different interests, check that your prose is clear and direct.

22 Check the summary (see p. 138).

23 Are all your revisions improvements? Is every word, letter, number and symbol in your manuscript legible?

24 Are all pages numbered and in their correct order?

25 Does the revised report read well and is it well balanced?

Preparing the typescript

When your investigation is complete the later drafts (with two carbon copies) should be typed. The original should be kept in a safe place; one copy can be used when you prepare the next draft; and the third copy can be given to someone else to read (see p. 44). There is a UNESCO guide for the preparation of scientific papers and author's abstracts (UNESCO, 1968). See also *Standards* p. 123. However, your typescript for an internal report, thesis, or research paper, must conform to the house rules, the regulations of the University, or the rules and notes for authors prepared by the editors of the journal to which your paper is to be submitted (see also Handbooks, p. 122). You should also study the format and style of a recent report, thesis, or paper, which conforms to these rules.

If possible, have your report typed by someone with experience of this kind of work. Otherwise, the typist should be told that normal office rules for typing correspondence, etc., do not apply, and should be given clear instructions.

Instructions required by the typist

1 Typescript required by .....

2 Use A4 paper (210 × 297 mm).

3 One top copy plus two carbons required.

4 Use clean type and a new black ribbon.

5 Type in double spacing on one side of the paper only.

6 Leave 40 mm margin on left; about 25 mm at right; and 25 mm at top and bottom of page.

7 Do not type part of a word on one line and part on the next;

and do not type a hyphen at the end or at the start of a line.

8　Number pages for typed report in the centre of the page at the bottom. Otherwise, for the printer, number each page at the top right hand corner.

9　For a printed report type the surname of the first author at the top left hand corner of each page.

10　Use a separate page for each table, with at least 40 mm margins. The number of the table and the heading should be immediately above the table. The heading should be underlined but it should not be followed by a stop, unless it is a sentence. For a printed report the tables should be at the end of the typescript before the legends to figures.

11　For a printed report, leave space in the typescript for the author to insert any mathematical expressions or chemical formulae.

12　Main headings (marked A in margin) should be centred at the top of a new sheet; sub-headings (marked B) shouldered and with a line to themselves; and minor headings (marked C) shouldered and followed on the same line by the next sentence.

13　Main headings and sub-headings are not to be followed by a stop, but a stop may be used after a minor heading to separate the heading from the next word on the same line.

14　All headings should be in lower case, except for the initial letter of each heading and for words that normally require capitals.

15　Do not indent the first paragraph after a heading. Indent all other paragraphs.

16　Do not underline headings. Underline only those words underlined in the manuscript (titles of books, names of journals, and scientific names of organisms).

17　*Contents* pages to be typed last when page numbers are known. For a printed report the page numbers will be added on the *Contents* pages by the printer.

18　For a printed report, the legends to the figures should be typed below the heading *Legends to Figures* and attached to the back of the typescript. A concise legend, if it is not a sentence, should not be followed by a stop.

Things to check in your typescript
Corrections, to be made in your typescript, should be marked on the bottom copy. Any marks on the top copy should be in soft black pencil so that the page does not have to be re-typed.

1 Does the report read well? Is it well balanced?
2 Are there any typing errors, or mistakes in spelling or grammar?
3 Are all dates, numbers, and mathematical and chemical formulae correct?
4 Are the references to tables and figures in the text numbered correctly?
5 Is the spelling of all scientific and proper names correct?
6 Check the wording and punctuation of all quotations and references against the original. If words are omitted from a quotation the gap should be indicated by three stops ... and anything added should be in [square brackets].
7 Are all references cited in the text up to date and in the list of references? Read the papers cited to make sure that you have taken the right meaning.
8 Are the headings to all tables and the legends to all figures adequate?
9 Is the source of any quotation, table or figure properly acknowledged, and where necessary has the written permission of the copyright owner (see p. 156) been obtained?
10 Have all diacritical marks (in quotations from other languages) or symbols been inserted in the right place?

If the report is to be printed occasional words may be corrected in ink between the lines (but not in the margin). Otherwise, pages should be retyped so that there are no extensive handwritten corrections.

If the report is to be printed, check that each folio (sheet of typescript or other copy) is numbered correctly (top right hand corner) and that the surname of the first author is also given (top left hand corner). If any folios are added later, the two preceding folios should be marked, for example: 29 (folios 30 a–c follow) and 30a (folio 30b follows) and the additional folios should be numbered 30c (folio 31 follows). If any folio is taken out, the preceding folio should be renumbered (for example, if folio 12 is removed, folio 11 is renumbered 11–12).

Preparing the index

The index for a typed report should not be prepared until the typescript is otherwise complete. The index for a printed report should be prepared from the page proofs.

Use one copy of your report (or proofs). Read the report, underlining in a conspicuous colour all words to be included in the index (topic words). Then look through the report, page by page, writing each word so underlined on a separate index card with the number of each page on which the word appears. Keep the index cards in alphabetical order (see BS 1749 *Alphabetical arrangement*) to facilitate the addition of page numbers. The index can be typed from these cards.

Sub-entries should be indented and arranged in alphabetical order below the relevant main entry. Unless the publisher specifies otherwise, each main entry and sub-entry should start on a new line; the first page number should be preceded by a comma and successive page numbers should be separated by commas. No punctuation is used at the end of a main entry or sub-entry. If the publisher specifies that sub-entries are to be run-on, separate them by means of semi-colons.

When an entry in the index refers to the main subject considered on successive pages, only the first and last of these pages are given, joined by a dash.

Cross references may be useful. Otherwise, the same page number should be given under different headings. *See also* references at the end of an entry may help the reader.

In typing the index (for a printer) use double or treble spacing, and wide margins, and keep a carbon copy. The pages with illustrations (or definitions) may be underlined (to be printed in italics) or underlined with a wavy line (to be printed in bold). A note should be included at the start of the index to explain that the pages with illustrations (or definitions) are printed in italics (or bold) as appropriate.

To avoid confusion with page numbers any dates in the index should be in parentheses (round brackets).

Recommendations for the preparation of indices are the subject of BS 3700 and ANSI: Z39.4 (ISO 999).

Preparing the typescript for the printer

Check the typescript (see p. 151) and illustrations (see p. 119). Make corrections and revisions in the typescript; not in the proofs. In the proofs even the addition or deletion of a comma involves respacing a whole line of print. Also, errors and omissions waste the time of editors and referees and give a poor impression of your work.

Where necessary, include marginal instructions for the printer. For example, explain any unusual symbols or Greek letters. Underline the words and symbols to be printed in italics. If anything that is correctly typed could be considered to be a mistake, write *set as typed* next to these words or letters. Use marginal letters to indicate main headings A ; sub-headings B ; and minor headings C . Indicate the position of each table and illustration by a marginal note in the text.

Photographs (see also p. 109) to be used in the preparation of plates should be black and white and they should normally be full plate or half plate. When several photographs are to be shown in the same plate, prepare a key for the printer to show the arrangement you require. Do not mount the photographs, unless required to do so by the publisher.

If only a part of the photograph is required, this part should be marked by a rectangle on a transparent overlay. Alternatively, prepare an enlargement from the relevant part of the negative.

Other information required by the printer should be marked lightly
on the reverse, preferably in the margin. Care is needed in writing
on the back of a photograph (or on an overlay) as lines may show
on the photograph and spoil the plate – as may an over-inked
rubber stamp on the back or pressure marks caused by paper clips.

When lettering, a scale, or other marks have to be inserted by
the printer, copies of the photographs with the necessary additions
should be provided or the additions should be printed on a transparent overlay, according to the requirements of the printer. If
any illustration is without letters or numbers, or some other clear
indication of its correct orientation, the word *top* should be written
lightly on the reverse (preferably in the margin).

Correspondence with an editor

If your report is to be published as a scientific paper, do not
publish prematurely. Do not submit a paper for publication unless
it is original and unless it includes new findings (and/or new ideas
supported by sufficient evidence). Remember that it is better to
publish in the most appropriate journal than to publish quickly
in a less appropriate journal. Do not submit the paper to more
than one journal at a time, and do not submit a typescript if it
has already been published or accepted for publication elsewhere.

Send your typescript to the editor at the address given in a recent
issue of the journal. The typescript (including the title page, text,
references, tables, and legends to figures) and the art-work, should
be kept flat with stiff cardboard and posted in one envelope. The
pages should be held together by a paper clip (not by a staple)
or they should be punched and threaded on a treasury tag. For
a long report one paper clip may be used for each part and then
an elastic band should be put around the whole typescript. If the
editor requires more than one copy, all copies should be sent in
the same envelope.

The editor will acknowledge receipt of your paper. Then there
will be a delay while he sends the paper to one or more referees
who are asked to judge the suitability of the paper for publication
in this journal. You can save yourself time and help the editor

and the referees if you consider the following questions before you submit your paper to an editor.

Check list for referees (and authors)
1 Is the paper suitable for publication in this journal?
2 Do you recommend publication of the paper (a) as it is, or (b) after revision?
3 Is the work reported original: has any part been published?
4 Is the work complete? Is it a contribution?
5 Are there any errors or faults of logic?
6 Is the paper clearly written? Are there ambiguities? Are any parts badly expressed? Are any parts superfluous? Are any points overemphasized or underemphasized? Is more explanation needed?
7 Does the paper conform to the rules of the journal?
8 Should all parts of the paper be published?
9 Is the title clear, concise and effective?
10 If key words are required, are these appropriate?
11 Is the abstract comprehensive and concise?
12 Are the methods sound? Are they described clearly and concisely?
13 Are the illustrations and tables properly prepared?
14 Are the conclusions supported by sufficient evidence?
15 Are all relevant references cited? Are any of those cited unnecessary?

If the editor wishes to accept your paper he is likely to suggest improvements. He speaks from his experience in publishing and his comments are based on the confidential reports of the referees. If you do not like what he says, do not reply immediately. Write a reasoned reply when you are ready to submit your revised paper. Both reviewers may be wrong but you should welcome their comments. Take the opportunity to think again, to correct any mistakes, to clarify any difficulties or ambiguous points, and to consider other revisions. In returning your typescript to the editor, you should say how it has been improved. If any suggestions have not been accepted you should say why. Responsibility for the type-script rests with you; just as responsibility for its acceptance or rejection rests with the editor.

Some journals receive many more papers for consideration than they are able to accept. Rejection, therefore, is not necessarily a reflection on the quality of your paper. Sometimes an editor suggests that another journal may be more appropriate. Some-times one editor rejects a paper, the importance of which is recognized by another editor. If the paper is rejected, however, you should read it again to see if it can be improved.

Copyright

Before reproducing copyright material you must obtain such per-mission as is required by law. For advice on copyright see the *Writers' and Artists' Yearbook* (A. & C. Black, London) but re-member that the law is not the same in all countries and if you are writing a review or a book you should consult your editor or publisher. If quotations are included for purposes of criticism or review, or if tables or illustrations are modified, the permission of the copyright holder may not be necessary. A proper acknowledgement of the source of the quotation (see p. 3) or of the original illustration upon which your figure is based (see Fig. 3, p. 29) may be all that is required. Copyright material should also be acknowledged (see p. 21 and Fig. 12, p. 112). For tables and illustrations copied *without modification* a fee may be required and the copyright holder may state how the acknowledgement is to be worded.

Anyone wishing to use copyright material should write both to the owner of the copyright and to the author (or publisher) of the work in which the material first appeared. In seeking permission to reproduce material, lines to be quoted should be identified by the title of the work, the date of publication (and the number of the edition and volume), the page number, and the numbers of the lines on which the quotation starts and ends, with the first few words and the last few words of the quotation. Illustrations and tables which are to be copied should be identified in a similar way but by their number and by the page number on which they appear.

Type three copies of this letter (1 + 2 carbons) with a statement below your signature in the form of a reply. This should say that permission to use the above mentioned material in the way described is granted. There should then be a space for the date and for a signature. Two copies of this letter should be sent, with a stamped addressed envelope, so that the copyright holder can return one signed copy and retain the other as a record.

Things to check in the proofs

Proofs are for checking and for the correction of printers' errors (in red ink), not for alteration. However, if you must make changes, any additions or deletions (in black or dark blue ink) should be matched by corresponding deletions or additions, of words or phrases of the same length (counting each letter and each space).

1 All notes for the printer, and any corrections, must be marked on the proofs, not on the typescript.
2 Corrections must be indicated clearly for the printer, in the right and left margins and with appropriate marks in the text. See BS 5261C and, for mathematical copy, BS 1219M; or see ANSI Z39.22, in list of Standards on p. 123. Words deleted should be crossed out by a horizontal line, and letters by a nearly vertical line. Any marginal comments or instructions for the printer, which are not to be set in type, should be encircled and preceded by the word PRINTER.
3 The questions asked by the printer, usually marked by a question mark in the margin, must be answered carefully.
4 Write the numbers on the contents page and in cross references in the text; and prepare the index.
5 Ask someone to read the typescript aloud while you check that the proofs are an accurate copy.
6 Read the proofs several times to check for printer's errors and for mistakes in spelling.
7 Check the accuracy of all dates, numbers and formulae.

8 Check the spelling of all scientific and proper names.

9 Check the wording and punctuation of all quotations and references against the original.

10 Check that the tables and figures are in the right place, that they have the right headings and legends, and that the numbers used in the text are correct.

11 Check the illustrations to ensure that they are a good copy of the original, that all lines are good, and that there are no extraneous marks.

Retain one copy of the corrected proofs and return one copy to the editor as quickly as possible.

Summary

How to prepare a report on an investigation or a paper for publication

This check list is a summary of the procedure recommended in this chapter together with relevant advice from other chapters. For a typed report, follow the stages marked by numbers below.

> Additional stages, necessary in preparing a paper for publication, are inset.

1 Keep full bibliographic details of every relevant reference consulted. See p. 128 *Reading*; and p. 139 *The list of references*.

2 Make a carbon copy of everything you write, and keep this in a safe place separate from the top copy.

3 Write the title and the *Introduction* and prepare a provisional list of *Contents*, with headings and sub-headings, before you start to work. Revise these as necessary during the investigation.

4 Keep a day-to-day record of your work in a laboratory notebook (see p. 8).

5 Write the *Materials and methods* section as soon as your procedure has been established and any initial difficulties have been overcome.

6 Prepare tables, pencil drawings, diagrams and graphs (with effective headings and legends on the same sheet at this stage) as each observation or experiment is completed.

See p. 99 *The use of tables*
p. 102 *The use of graphs and diagrams*
p. 107 *Illustrations contribute to clarity*
p. 107 *Writing the legend*

7 Prepare notes on each observation or experiment as it is completed.

See p. 9 *Writing helps you to observe*
p. 11 *Writing helps you to think*

8 List points arising from your work and from relevant work by other people so that you will remember these when you are ready to write the *Discussion*.

9 When your work is complete prepare a detailed topic outline.
   - Consider whether your work should be published.
     If there are new findings, should your report be published as a whole or in part? Should it be published as one or more papers?
   - If necessary, revise the topic outline(s).
   - Consider carefully which journal(s) your paper(s) would fit into most appropriately.
   - Read the *Instructions to Authors* for the journal.
   - If necessary, revise the topic outline to ensure that your paper conforms to these instructions.

10 Write the first complete draft.

See p. 42 *Writing*
p. 131 *The parts of a research report*
p. 146 *Preparing the manuscript*

11 Revise the first complete draft

See p. 43 *Revising*
p. 147 *Things to check in your manuscript*

12 Ask two people to read the second draft (see p. 44); and then revise the paper in the light of their suggestions.
   - Check that publication is acceptable to your employer and that nothing to be published is classified as confi-

dential or secret. See also p. 45 on patents.

13 Obtain written permission to use any copyright material (see p. 156).

14 Read all the references cited in the text to make sure that the work of others is correctly represented and that the bibliographical details on your index cards are correct and complete. See p. 137 *The Discussion*; and p. 139 *The list of references.*

15 Make sure that everything in the manuscript is in the right place.
    - For a printed report the tables should be placed after the list of *References* and not in the appropriate part of the text.

16 Give appropriate instructions to the typist.
    See p. 149 *Preparing the typescript*

17 Have the list of *References* typed from the index cards.

18 Prepare the illustrations in ink (see p. 114).
    - Prepare the *Legends to figures* (see p. 117).
    - Have these typed and add them at the end of the typescript (after the tables).

19 Check the typescript (see p. 151). Make corrections on a carbon copy, so that the typist can correct the top copy.
    - Check that the typescript meets all the requirements listed in the journal's *Instructions to Authors.*
    - Minor corrections, additions or deletions should be marked clearly in ink on all copies of the typescript.

20 Obtain clearance for the corrected typescript from your supervisor, head of department, or employer, as appropriate.
    - Mark the typescript for the printer, and keep a copy marked in the same way.
    - Send the number of copies required by the editor of the journal, with a short covering letter, to the editor. See p. 154 *Correspondence with an editor*
    - Correct the proofs (see p. 157); and return one copy of the corrected proofs to the editor with your order for reprints. Keep one corrected copy of the proofs.

14

Talking about science

The expression *reading a paper* is misleading, for speech is not the same as writing (see p. 73). If your talk is to be published it is best to prepare the typescript for publication and then to speak from notes. Listeners cannot assimilate all the detail that goes properly into a written report but which has no place in a talk.

Preparing a talk

Define the purpose of your talk. If you have been asked to talk on a particular subject, keep to your terms of reference. Decide on a limited number of important points. Deal with these, in your talk, in a logical sequence: Why was the work done? How? What did you find? Inexperienced speakers usually try to make too many points and to support their argument with too much detail.

Get to the point quickly at the start of each aspect of your talk. Read your talk aloud, to yourself, to check that you have not written something that you would not say. If possible, try out your talk on your colleagues.

Repetition is usually undesirable in writing because the same words can be read again but a lecturer must communicate the essentials. One way to do this is by repetition: say what the talk is about; say what you have to say; briefly rephrase what you have said to ensure that everyone understands each stage of your talk; and summarize your main points towards the end of your talk so that they lead to your conclusions.

People will listen most carefully to, and will remember best, what you say in the first fifteen minutes; and thirty minutes with one person talking is enough for any listener. A well planned talk may follow this plan: brief introduction – main points – elaboration and visual aids – conclusions – discussion.

Timing

Relate the number of main points to be made to the time available, the needs of the audience, and the time needed to introduce the subject and to round off your talk effectively. The time devoted to the use of a blackboard or other visual aids must be deducted from the time available for speaking.

Some speakers underline key words in their lecture notes and mark the margin to indicate where they should be at different times. Others use index cards; with one card for the introduction, one for each main point, and one for the conclusions.

Leave time, after your talk, for questions and discussion. Remain at the front of the room, facing your audience. Make notes while other people are speaking so that you can give a complete reply. Because the time for discussion is short, all questions, answers and comments should be brief.

Nobody minds if your talk ends a few minutes early but do not speak for too long. Also, chairmen should not encroach on the time allocated to a speaker. Chairmen will find it easier to keep speakers to time if they allow them to start on time. Chairmen should also talk to each speaker privately before the talk rather than by an interruption shortly before the finishing time or after this when the speaker has already spoken for too long.

Complete your final arrangements shortly before you talk. Make sure you know how to use any switches or projection equipment; or brief the projectionist. Open a few windows. Most people will be sleepy if they are made to sit in an overheated and poorly ventilated room.

Using a blackboard

The blackboard is more often misused than properly used. Perhaps this is why, with other visual aids available, some people speak disparagingly of talk and chalk.

1 Yellow or white chalk is best on a green or black board.

2 Everyone is further away from the board than you are, so write in large clear block capitals.

3 Use diagrams; not drawings. Use diagrams that can be built in stages. Plan such diagrams so that you can leave space for the things which are to be added later.

4 When you are writing or drawing – stop talking. Try not to obscure anyone's view. Turn to look at your audience if you wish to speak, and to make sure that they are looking at what you wish them to see.

5 People may need time to study a diagram quietly, without the distracting effect of your voice.

6 Spell any word that may be new to some people in your audience, and any word that may be pronounced differently by people who do not normally speak English.

7 Keep the blackboard clean (Fig. 17). Do not let people look at one thing when you are trying to interest them in something else. Let any words and diagrams stand out.

Using slides

In planning your lecture decide when to show any slides. It is disturbing to everyone if you switch lights off and on repeatedly. But if the lights are off all the time you cannot look at your audience – to hold their attention and to see their reactions. Try to show the slides in one batch. Talk first and then show your slides; or use the slides to provide a break in a long lecture or to separate the body of your talk from the summary and conclusions. What is best for one talk may be inappropriate for another.

Many people prefer to use slides rather than a blackboard for tables and diagrams. These save time if you have a lot to say but a common result is that lecturers say too much and show too many slides, in an attempt to present more information than people can assimilate in the time.

1 Do not show too many slides.

2 Sit at the back of a large lecture theatre while your slides are projected so that you can check them for clarity.

3 Do not show a slide if it includes too much detail or anything that is not relevant to your talk. Use one slide to convey one message and make the message brief, clear and simple so that it can be understood quickly.

4 Do not show a table if it has too many numbers or if the numbers are so small that some people cannot read them.

5 Arrange the slides in the same order as your notes.

6 Make sure that each slide is the right way up and the right way round.

7 Give the audience time to look at each slide and then help them to interpret the slide.

8 Remove each slide as soon as it has served its purpose, so that it is not displayed while you are trying to interest your audience in something else.

Tables and illustrations from a published paper will probably include more detail than is acceptable in a slide for projection (unless they were prepared with both uses in mind: see p. 123 (ANSI Y15.1).

Slides for projection should have a concise heading, not a legend. Any further explanation (of symbols, for example) should be part of the illustration.

Tables that are to be made into slides can be prepared with a typewriter. Use a new black ribbon and clean type. Type in double spacing. In pica (10 letters to 2.5 cm) type in a rectangle 12 cm × 12 cm, and in elite (12 letters to 2.5 cm) in a rectangle 10 cm × 6.5 cm. The title should be one line of about seven words, and there should be up to four columns with up to ten items in each.

Delivery

Do not begin by saying that you are not really qualified to speak on this subject. Decide about this before you agree to talk.

Begin well – to capture the attention of everyone present. Some speakers use a joke to put people at ease. Others cause embarrassment when people feel obliged to laugh. While jokes are not out of place in some lectures, they are best avoided in a report on a scientific investigation.

Many lecturers have favourite words ... *sort of*, *like*, *er*, *I mean* ... which they repeat so often that they distract the listeners' attention from what is being said. Unwanted words and phrases such as these, which give the speaker time for thought, may be a sign of inadequate preparation. Other expressions – *you see*, *you know*, *all right*, and *if you follow me* – are attempts at confirmation.

Your vocabulary, approach, and speed of delivery, should match the interests and the size of your audience. The lecture to a large audience is formal. The seminar is informal and more time is needed for questions, clarification and discussion.

Some speakers use carefully considered gestures to good effect. Some juggle with chalk or with a pointer, and so distract people's attention. Others behave as if they are on sentry duty. Others never move and give too rigid a performance.

Speak so that everyone can hear every word; but do not use a microphone unless poor acoustics make this necessary. Stop talking whenever you face away from your audience. Look at your audience to make sure you have their attention. Let everyone see your expressions. By looking around the room try to be aware of those people who understand and of those who require further explanation.

Bring your talk to an effective conclusion. Leave people to reflect on your last words. If you talk for too long they may remember only that you did not know when to stop.

References

Allbutt, T. C. (1923). *Notes on the Composition of Scientific Papers.* 3rd edition. London: Macmillan.

Almack, J. C. (1930). *Research and Thesis Writing.* Boston: Houghton Mifflin Co.

Baker, J. R. (1955). *Nature* **176**, 851–2.

Beveridge, W. I. B. (1968). *The Art of Scientific Investigation.* 3rd edition, London: Heinemann.

Drucker, P. F. (1952). How to be an employee. *Fortune* **45** (5), 126–7. Reprinted in Emberger and Hall (1955) p. 297.

Emberger, M. R. and Hall, M. R. (1955). *Scientific Writing.* New York: Harcourt Brace.

Flesch, R. F. (1962). *The Art of Plain Talk.* London and New York: Collier–Macmillan.

Flood, W. E. (1957). *The Problem of Vocabulary in the Popularisation of Science.* University of Birmingham; Edinburgh: Oliver & Boyd.

Fowler, H. W. (1974). *A Dictionary of Modern English Usage.* 2nd edition revised by Sir Ernest Gowers. Oxford: Clarendon Press.

Gill, R. S. (1954). *Science* **119**, 3A.

Gowers, E. (1973). *The Complete Plain Words.* 2nd edition revised by Sir Bruce Frazer. London: HMSO.

Graham-Campbell, D. J. (1953). *Writing English.* London: Edward Arnold.

Graves, R. and Hodge, A. (1947). *The Reader Over Your Shoulder.* London: Cape; New York: Macmillan.

Grogan, D. J. (1973). *Science & Technology: an introduction to the literature.* 2nd edition. London: Bingley.

Harvey, A. P. (Ed.) (1969). *Directory of Scientific Directories.* New York: Guernsey: Hodgson; New York: Int. Publ. Service.

Henn, T. R. (1960). *Science in Writing.* London: Harrap.

HMSO (1972). *Education: A Framework for Expansion.* Cmnd. 5174, London: HMSO.

HMSO (1975). *A Language for Life.* London: HMSO.

Kapp, R. O. (1973). *The Presentation of Technical Information.* 2nd edition revised by Alan Isaacs. London: Constable.

McCartney, E. S. (1953). *Recurrent Maladies in Scholarly Writing.* Ann Arbor: University of Michigan Press.

McClelland, E. H. (1943). *J. Chem. Educ.* **20**, 546–53.

McIntyre, I. (Ed.) (1975). *Words.* London: BBC.

Napley, D. (1975). *The Technique of Persuasion.* 2nd edition. London: Sweet & Maxwell.

Orwell, G. (1937). *The Road to Wigan Pier.* London: Gollancz; New York: Harcourt Brace.

Orwell, G. (1950). Politics and the English language. In: *Shooting an Elephant and Other Essays.* London: Secker & Warburg; New York: Harcourt Brace.

Partridge, E. (1953). *You Have a Point There.* London: Hamish Hamilton.

Partridge, E. (1965). *Usage and Abusage; a guide to good English.* 8th edition. London: Hamish Hamilton; New York: British Book Centre.

Potter, S. (1969). *Our Language.* Harmondsworth & New York: Penguin Books.

Quiller-Couch, A. (1916). *On the Art of Writing.* Cambridge University Press.

Reeder, W. G. (1925). *How to Write a Thesis.* Bloomington, Illinois: Public School Publ. Co.

Rivet, A. L. F. (1976). *Times Higher Education Supplement* **222**, 5–00.

Royal Society (1974). *General Notes on the Preparation of Scientific Papers.* 3rd ed. Royal Society, London, 31pp.

Sampson, G. (1925). *English for the English.* 2nd edition. Cambridge: University Press.

Smith, G. O. (1922). Plain geology. *Econ. Geol.*, **17**, 34–9.

Strong, L. A. G. (1951). *English for Pleasure.* London: Methuen.

Thornton, G. H. and Baron, K. (1938). *Teach Yourself Good English.* London: English Universities Press.

Tichy, H. J. (1966). *Effective Writing for Engineers – Managers – Scientists.* New York & London: Wiley.

UNESCO (1968). Guide for the preparation of scientific papers for publication *and* Guide for the preparation of authors' abstracts for publication. UNESCO, SC/MD/5. Aug. 1968, 7pp.

Vallins, G. H. (1960). *The Best English.* London: Andre Deutsch.

Vallins, G. H. (1964). *Good English: How to Write It.* London: André Deutsch, Washington: Academic Press.

Warner, G. T. (1915). *On the Writing of English.* London & Glasgow: Blackie.

Index